IMAGES
of America

COAL IN CAMPBELL COUNTY

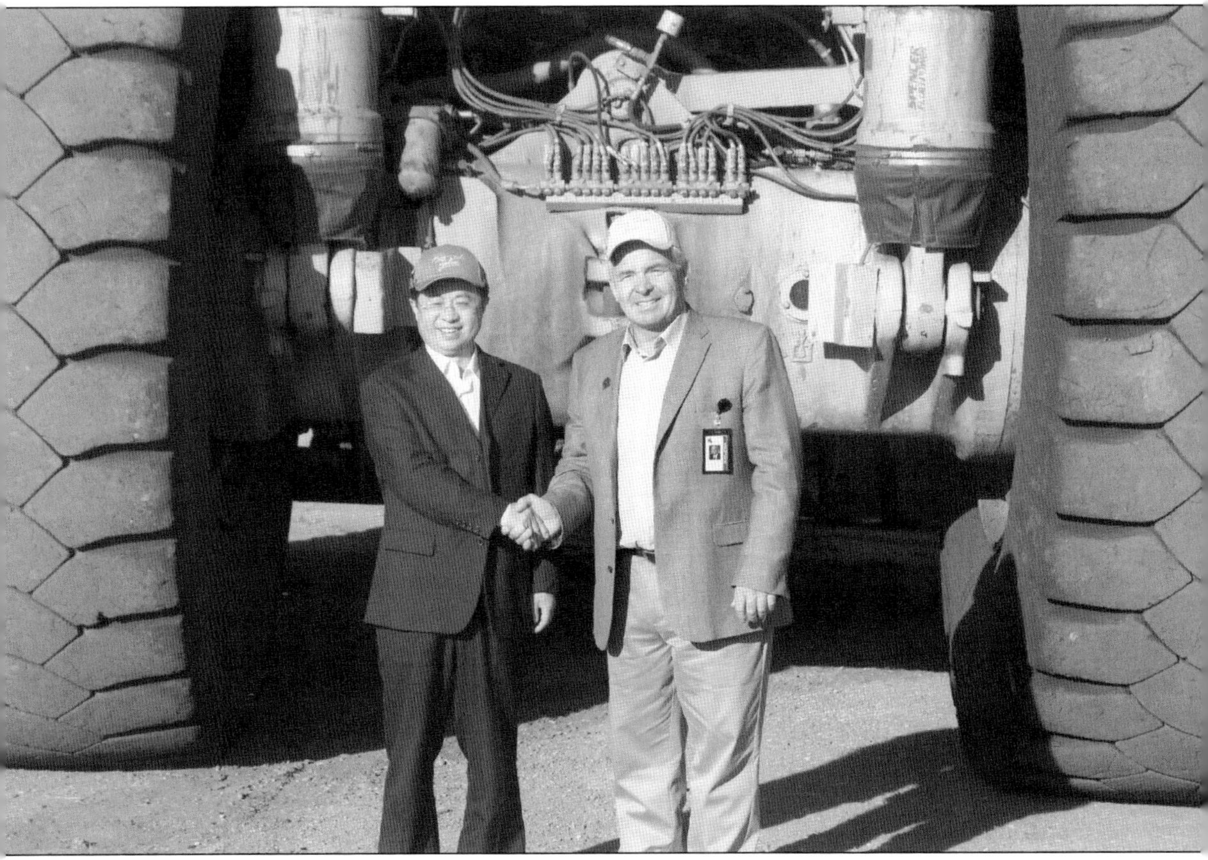

Heavy regulations and an uncertain future for coal in the United States have led to an industry interest in exporting coal to China. Gillette mayor Tom Murphy, right, and Mayor Lu Zhiyuan from the city of Yulin, China, have been working for almost a year to create a relationship wherein Campbell County can increase production to meet China's needs, and China can accommodate its need to fuel its power plants and steel production. (Courtesy of the City of Gillette.)

On the Cover: The Civilian Conservation Corps in Campbell County, Wyoming, was the only camp of its kind developed to put out underground coal fires, many of which had been burning for years. Some fires called for the use of large steam shovels, but in this mine opening, hand tools and wheelbarrows made more sense. The prairie homesteaders learned that where there are coal fires, there is more coal that can be used as precious fuel.

IMAGES
of America

COAL IN CAMPBELL COUNTY

Mary Kelley with photographs from
Campbell County Rockpile Museum

ARCADIA
PUBLISHING

ISBN 978-1-4671-3017-2

Published by Arcadia Publishing
Charleston, South Carolina

Printed in the United States of America

Library of Congress Control Number: 2013934480

For all general information, please contact Arcadia Publishing:
Telephone 843-853-2070
Fax 843-853-0044
E-mail sales@arcadiapublishing.com
For customer service and orders:
Toll-Free 1-888-313-2665

Visit us on the Internet at www.arcadiapublishing.com

This book is dedicated to my husband, Mike, and our son Sean. The coal industry in Campbell County has provided a wonderful life for us.

Contents

ACKNOWLEDGMENTS

As with my first book, *Gillette*, the Rockpile Museum staff was my closest partner in the production of this book. While Robert Henning, registrar, was working on a new coal mining exhibit for the museum, he and I were also working on this book. Between our individual research, we came up with some new images and interesting anecdotes for both projects. The Rockpile Museum Board of Directors readily agreed to the idea of the new book, reflecting upon the success of the first.

While Robert and I were able to use some photographs from the museum's collection, the coal companies and service industries owned the real collections. Without their cooperation, most of these photographs would not be in this book. Some people provided a wealth of images; some provided one or two. Sandie Jones helped me fill in some blanks in the Peabody chapter. Peggy Keck and Joe Lynde helped with the first chapter on small, family mines. Other friends who provided photographs included Keith Fare, Alice Rohde, Lynn Stodola, Norman Grams, Donna Crippen, and Randy Thomas. A trip to Deadwood and Lead, South Dakota, resulted in photographs from the Homestake Adams Research and Cultural Center, and the Black Hills Mining Museum. I could have stayed all day at any of these places and looked at all the old photographs. I would also like to thank the American Heritage Center, the National Archives, and Nathan Doerr at the Sheridan County Museum for making a career of having great photographs available for people like me to borrow for our unique projects.

For the coal companies, I'd like to thank the following people:

Alpha: Ken Kehn, Mike Hopper, Carol Jandreau, Tommie Butler, Chuck Sumner
Cloud Peak: Mike Dunham, Molly Nicholson, Kay French, Bob Green, Mike Kelley, Bill Wallace
Peabody: Charlene Murdock, Chris Taylor, Mary Lou Butler, Sandie Jones
Arch: Bonnie Clemetson, Bob Stowe, Larry Curtis
Basin Electric/Dry Fork Mine: Curt Pearson, John Barnes
Kiewit: Debbie Noonan, Herb Pridgeon
Wyodak: Kurt Triscori, Dee Likewise
Lost Mines: Joe Lynde, Marilyn Drew

At the service industries, I'd like to thank the following people:

L&H: Brittney Thomas, Mike Wandler
Joy Global: Lana Randen, Roger Connett, Josh Tompkins
Interstate Power: Charlie Kiesel
Wyoming Machinery: Waylon Strohschein, Matt Cross
Cummins: Carrie Wilkerson, Sarah Fenneman
Railroad: Will Cunningham

Last but not least, it is important to recognize the *Gillette News Record* and its predecessors for recording the events of the day so that later generations would know the history of Campbell County. All photographs are courtesy of the Rockpile Museum, unless otherwise noted.

Introduction

Before 1972, Campbell County could boast of one commercial coal mine—Wyodak, which had been in existence since 1923. In 2012, a total of 12 mines operated in Campbell County, making Campbell County by far the largest coal producing county in the state of Wyoming. The mines in Campbell County produce coal from the Powder River Basin, 1 of 10 coal fields in Wyoming. The Powder River Basin covers 120 miles east to west and 200 miles north to south. The majority of the field lies in Wyoming, but it extends north into Montana. It is the single largest source of coal mined in the United States and contains one of the largest deposits of coal in the world.

The coal beds in northeastern Wyoming began to form about 60 million years ago. Where the Powder River Basin is now was a flat, sea-level plain, muddy from a retreating sea that had covered Wyoming for 40 million years. The land began to rise, causing the sea to drain away. Over the next 14 million years, the Powder River Basin rose to its present 3,000 feet, with the Big Horn Mountains and the Black Hills rising thousands of feet higher. The climate was subtropical, averaging 80 degrees Fahrenheit, with 120 inches of rainfall annually. Palm trees and crocodiles were common in the area. The mountains rose in pulses around the basin. Each pulse buried the swamps with peat sediment. The sediment turned into rock and coal over time. As the climate became drier and cooler, the swamps disappeared, and the basin filled, burying the peat under thousands of feet of rock. This caused the temperature of the peat to rise, changing it into coal. The last few million years have been a time of erosion, allowing the coal to be pushed almost to the surface, perfect for open-pit mining.

Western coal, such as that found in Campbell County, and eastern coal are significantly different due to different swamp environments, geologic age and local geology. To compare:

Western Coal	Eastern Coal
Subbituminous	Bituminous
Lower in Btu	Higher in Btu
Lower in ash	Higher in ash
Lower in sulfur	Higher in sulfur
Thicker seams	Thinner seams
Surface mining	Underground mining (some surface)

Bituminous coal is older and has a range of 10,500 to 15,000 British thermal units (Btus) per pound. Bituminous coal is found in Pennsylvania, West Virginia, and Kentucky. Subbituminous coal is younger and has a range of 8,300 to 11,500 Btus per pound. Subbituminous coal is found primarily in Wyoming.

At the risk of getting too technical, it is important to understand the different types of coal and the significant coal resources found in Campbell County. The evolution of the coal industry and the success of mining in the Powder River Basin are directly tied to the qualities of subbituminous coal and its proximity to the surface.

In 1921, Homestake Mining Company in Lead, South Dakota, purchased the property next to the Peerless underground mine in eastern Campbell County, and named it Wyodak. The Wyodak surface mine was developed and began production in 1925, supplying coal to the Homestake gold mining company in South Dakota. For the next 48 years, Wyodak was the only commercially operated coal mine in Campbell County. Some small, family-owned mines still operated, but not as businesses. Then, in 1972, AMAX started shipping coal from the Belle Ayr Mine, the first of many mines developed in the 1970s in the Powder River Basin.

I worked at AMAX Coal Company from 1974 to 1987—not in Campbell County, but at the national coal headquarters in Indianapolis, Indiana. While working there, I learned that the company began as Ayrshire Collieries. From 1969 to 1993, the name was AMAX. In 1993, the company became Cyprus. In 1999, it was sold to RAG. In 2007, the name was Foundation Coal. In 2009, it became Alpha Natural Resources. In this case, the two mines—Belle Ayr and Eagle Butte—remained as assets of the company, regardless of the owner and the name. Recording these changes—to all coal companies, not just AMAX—was why I wanted to write this book.

One

The Lost Mines

Northeastern Wyoming is not an area populated by trees. Therefore, early pioneers had to find other sources of fuel. Fortunately, Campbell County has an abundance of coal near the surface. Many pioneers made forays to cliffs and creek banks with a wagon to load up on coal for their homes. When the coal close to the surface was used up, enterprising settlers sunk shafts to access the deeper coal. Many of these mines were not safe, and tragedy often occurred when a roof caved in. Most men had no experience in mining. As early as March 1925, the state mine inspector was responsible for visiting the small local mines and ensuring they were in good condition. In a recent listing of Campbell County Mines, 68 mine names appeared. Some such as Jeanette Mine, Betty No. 12 Mine, and Syl Del Mine are not well known. In fact, no one in the industry could point out where they were located. Others such as the Ditto Mine, Christensen Brothers Mine, and Van Bugenum No. 2 Mine give clues to their location in their names. Some early mines had a small town named for them. The town of Wyodak was the most well-known, but the Shields Mine also had a small town named after it. The towns were intended to be the residence of the mine employees and often included a company store, school, and post office. The Peerless Mine was also owned by Mark Shields. It was built in the Wyodak coal seam, which ranges from 25 to 190 feet in thickness. The Peerless shut down in March 1923 after the death of Shields. In December 1923, the Kirby Mutual Coal Company bought the mine, with production starting again in January 1924. Mine inspector records showed the mine idle during 1925, 1926, and 1927, when the mine name drops off the mine inspector books. It is believed the mine closed for good sometime in 1927.

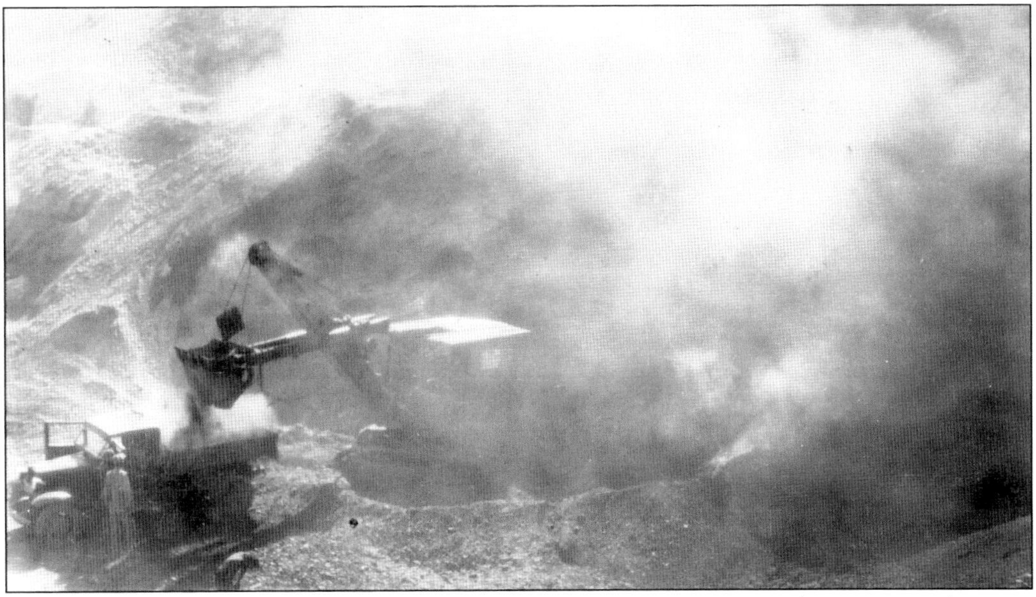

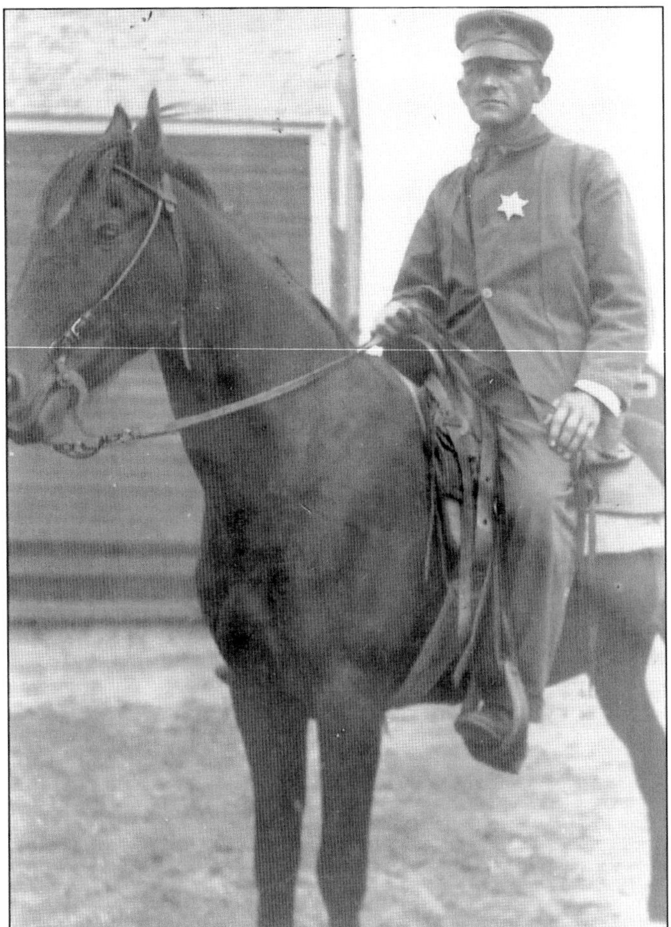

Many early mines caught fire, like this family mine on S.H. Rice's property. This was exactly the type of fire the Civilian Conservation Corps worked on in the mid-1930s in Campbell County. CCC workers use a steam shovel and dump truck to dig out the fire and spread the burning coal until the fire was extinguished.

Horace Ray Brooke, shown here when he was town marshal in 1911, died instantly in a coal mine cave-in on September 16, 1915. He and Charlie Broadhurst were digging coal in a mine on government land that was used by many residents but had never been secured with timber inside. A 50-foot wall of limestone, slate, and shale covered Brooke, killing him; Broadhurst was not injured.

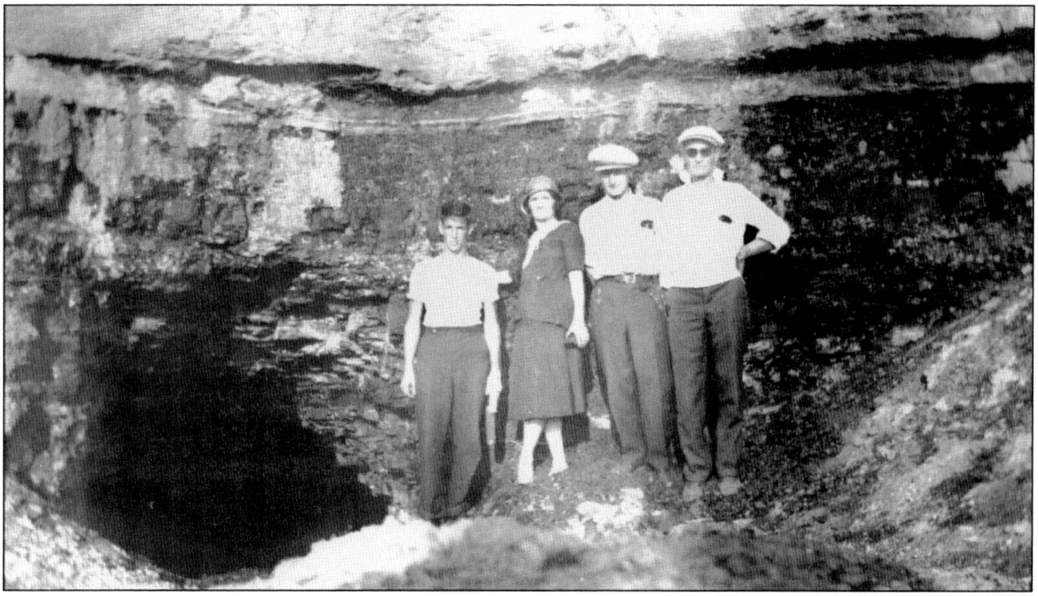

The coal-mine where Brooke died was also where George Larabee Vanderheyden was injured in a cave-in in 1926. Shown here in the 1930s are, from left to right, Marvin Ripley, Goldie Ripley, Frank Ripley, and Max Ripley, all relatives of Mr. Vanderheyden visiting from Iowa. (Courtesy of John Vanderheyden.)

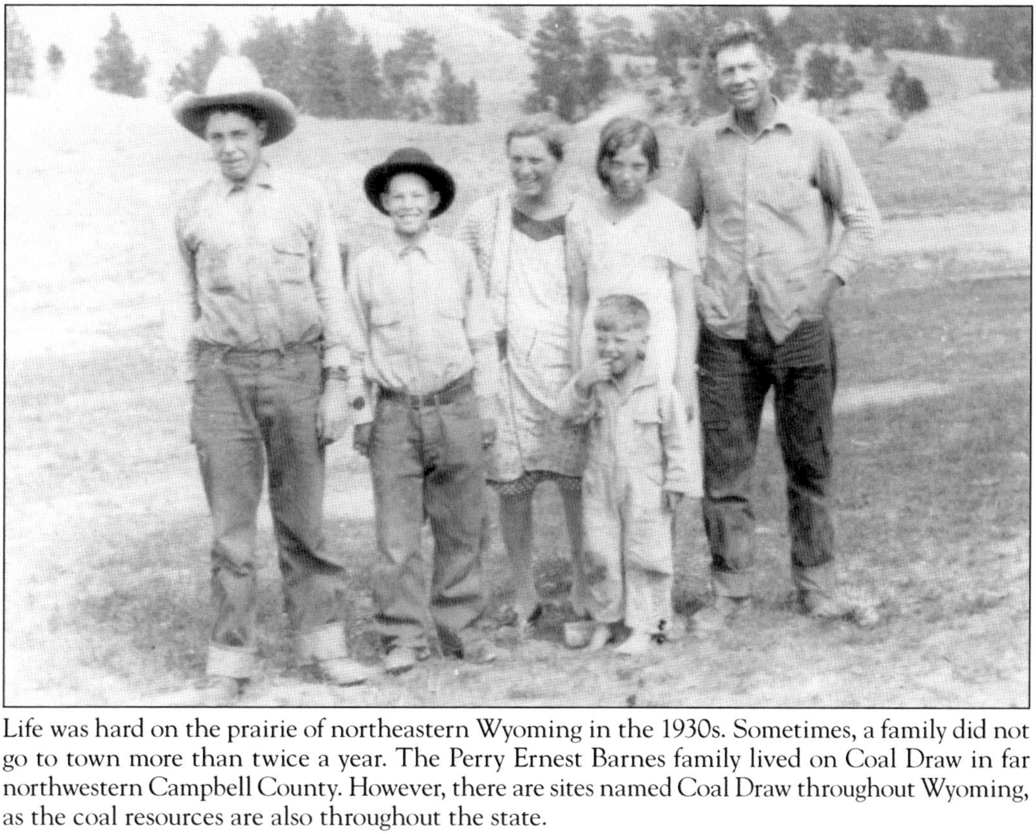

Life was hard on the prairie of northeastern Wyoming in the 1930s. Sometimes, a family did not go to town more than twice a year. The Perry Ernest Barnes family lived on Coal Draw in far northwestern Campbell County. However, there are sites named Coal Draw throughout Wyoming, as the coal resources are also throughout the state.

The H

Peerle

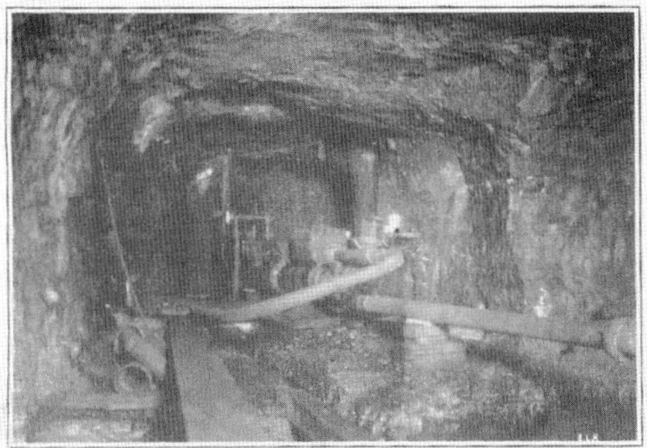

From the Thickest Vein in the United States

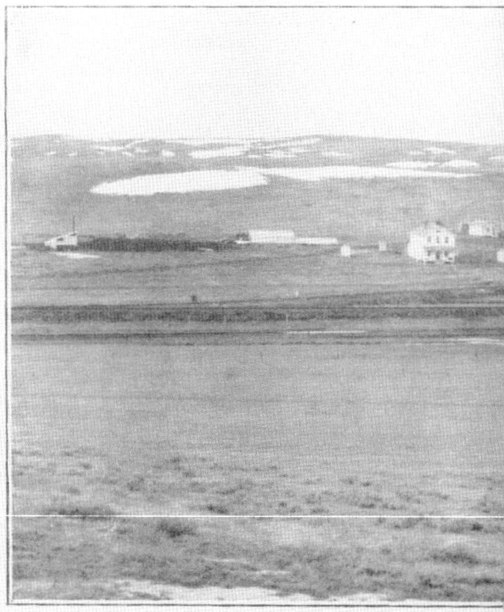

PEER

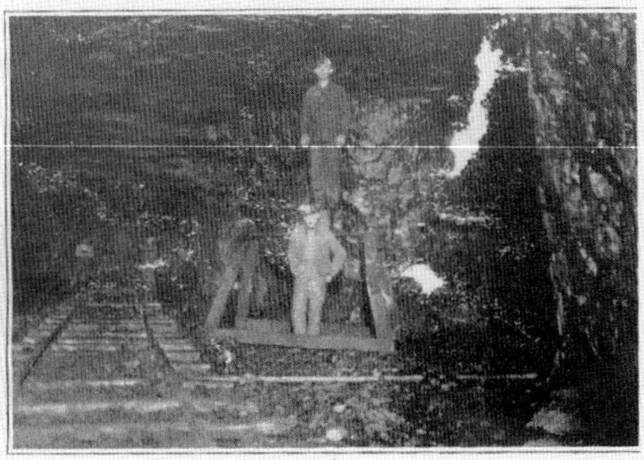

A Small Portion of Our 90 Foot Vein

Mined and S

KIRBY MUT

PEERLESS MINE

In 1874, gold was discovered in the Black Hills on French Creek by Ross and McKay, two miners attached to General Custer's military expedition. In 1876, the same year General Custer made his last stand at Little Big Horn, brothers Fred and Moses Manuel, along with Hank Harney, staked the Homestake gold claim in what is now known as the Open Cut in Lead, South Dakota. The first gold shipment, valued at $38,789.75, was shipped on August 17, 1878. It would be 40 more

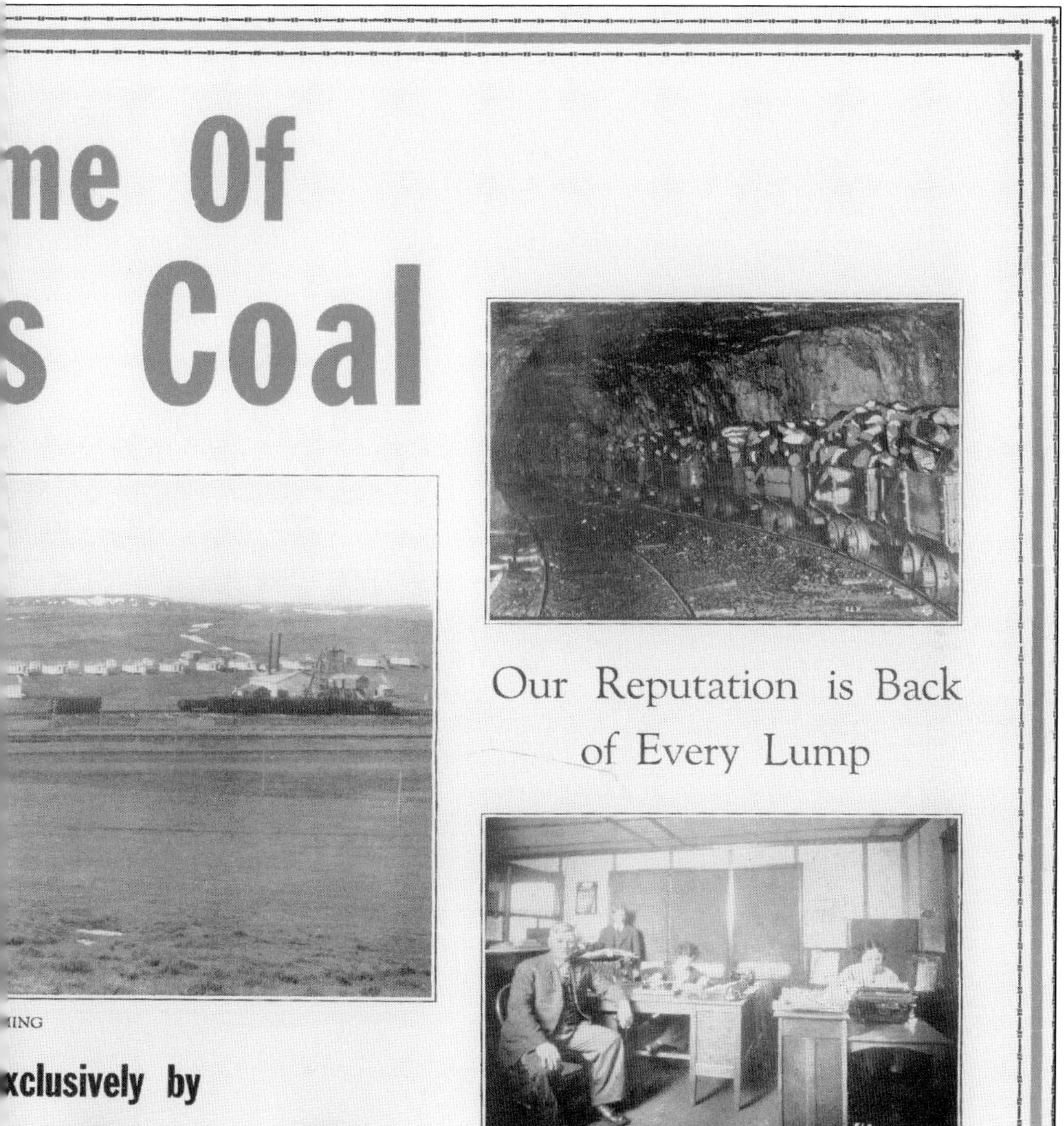

years until the Peerless Mine would begin operations in 1918. By 1925, Peerless would be idle, and Wyodak would be shipping coal to Lead. The small company town named Peerless was created on mine property, providing housing for mine workers and their families. This circular was distributed in 1923, the only year the mine operated when Kirby Mutual owned it.

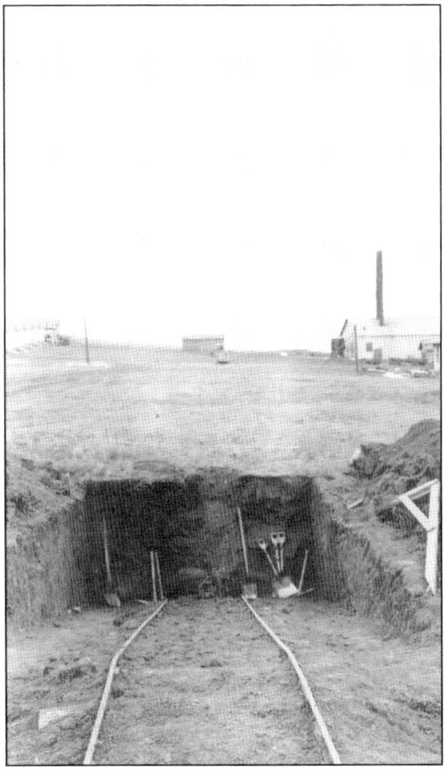

The Peerless Mine sunk a shaft 117 feet to access the coal, but also created a slope to haul the coal out of the mine. Two railroad tracks were laid leading into the mine, so railcars could be used to bring coal to the surface. Other mine buildings included a store, boardinghouse, office building, and tipple—a structure that can store, size, crush, or load coal into train cars.

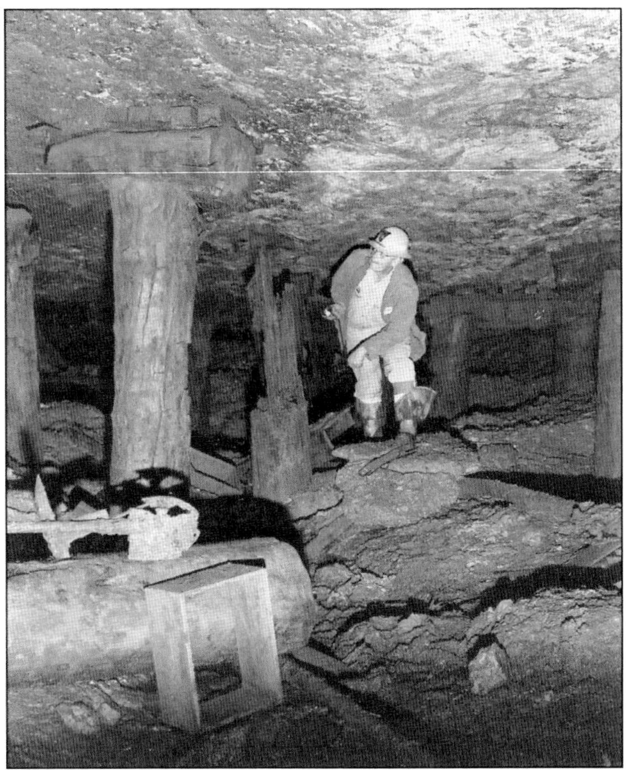

This photograph shows the underground Homestake Mine in Lead, South Dakota. Underground mining conditions in the 1920s were dark, dangerous, and uncomfortable. This mine is representative of the conditions in the Peerless Mine. Note the wooden pillars used to stabilize the roof. (Courtesy of Homestake Adams Research and Cultural Center.)

Mark Shields was the owner of the Peerless Coal Company and president of the Bank of Gillette. His downfall was investing bank funds in the marketing and promotion of the coal mine. When his actions were discovered, he killed himself on March 9, 1923. Shortly thereafter, the bank failed and the mine closed.

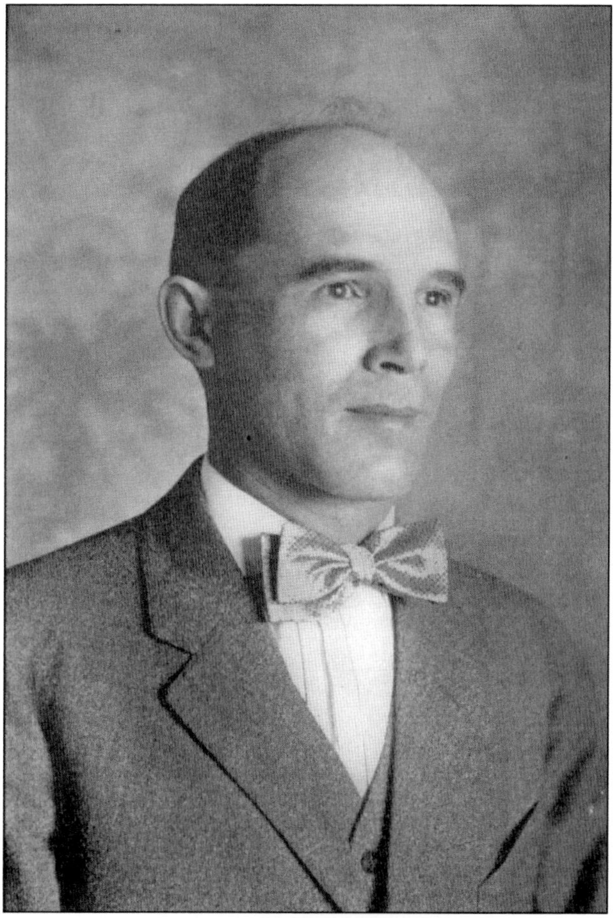

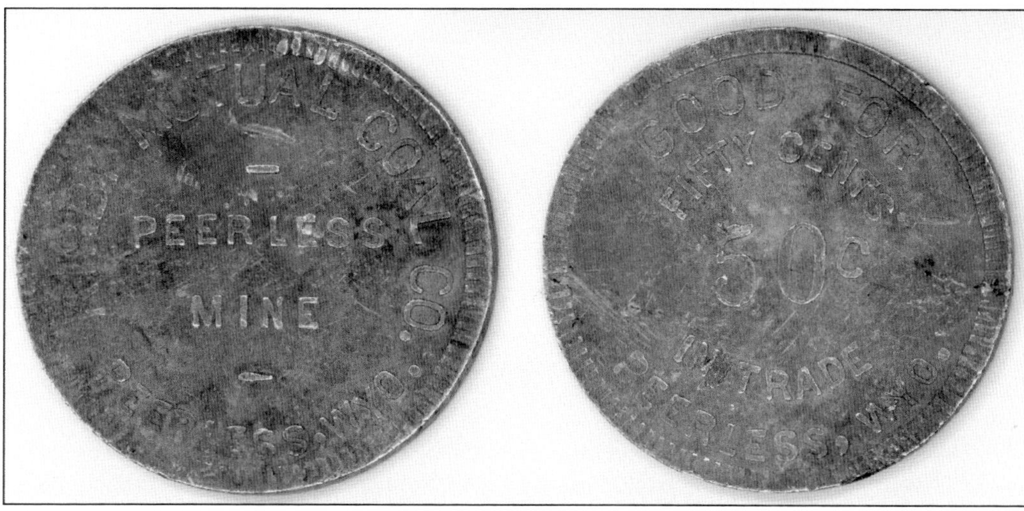

This coin reads "Kirby Mutual Coal Company, Peerless Mine, Peerless, WYO" on one side and "Good for Fifty Cents in Trade, Peerless, WYO" on the other side. It is thought to be legal tender that the mine workers could use at the company store in the town of Peerless, Wyoming. In the 1920s, a 50¢ coin could buy one pound of coffee or five pounds of flour.

DAILY TIME SHEET

Kirby Mutual Coal Company

Basin, Wyoming

PEERLESS MINE — Oct 20, 1924 DATE

CHECK NO.	NAME	RATE	HOURS	DISTRIBUTION	ACCT. NO.	DETAIL
1	Wm. Addy	98¾	8	Firing	$34a^2$	7.90
2	Wm. Soloman	99	3½	Boiler Coal	$34a^2$	3.46
7	R. Sandberg	98¾	7½	Tipple Scale	218a	7.41
8	Bert Cullivan	99 97¼	4 4	Loading Pumping	$12a^2$ $23a'$	3.96 3.89
9	Ralph Snyder	94	8	Firing	$34a^2$	7.52
10	Jesse Herx	94	8	Firing	$34a^2$	7.52
19	Joe Vitaley	99	8	Scale	218a	7.92
24	Jake Kavka	99	8	Scale	218a	7.92
27	Geo. Kluz	97¼	8	Boiler Flues	134B	7.78
34	Geo. Secora	99	8	Scale	218a	7.92
35	Preston Soloman	88	8	Boiler Coal	$34a^2$	7.04
40	Louis Zuige	99	8	Sup't	$36a'$	7.92
44	Adam Byrtus	99	3½	Boiler Coal	$34a^2$	3.46
45	John Rubich	99	8	Scale	218a	7.92
			102½			99.54

The Kirby Mutual Coal Company was based in Basin, Wyoming, about200 miles west of Gillette. This Daily Time Sheet, dated October 20, 1924, lists some of the mine employees, their rate of pay, hours worked that day, and the job they performed. William Addy received 98.75¢ for each of the eight hours he worked in "firing," resulting in a daily pay of $7.90.

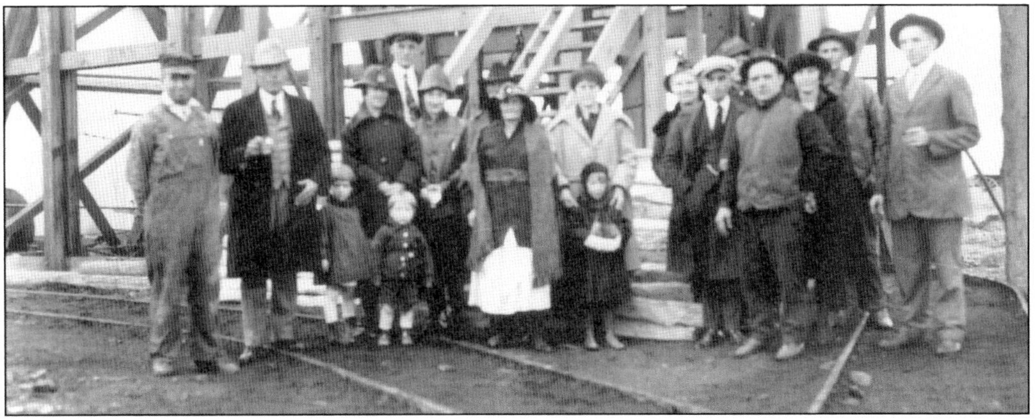

Fifteen visitors posed for a photograph at the Peerless tipple on a Sunday afternoon in 1923. Little did they know, many years down the road, that coal and its by-products would be used to make coke used in steel blast furnaces, soda water, acetylene, synthetic rubber, charcoal briquettes, and artificial silk.

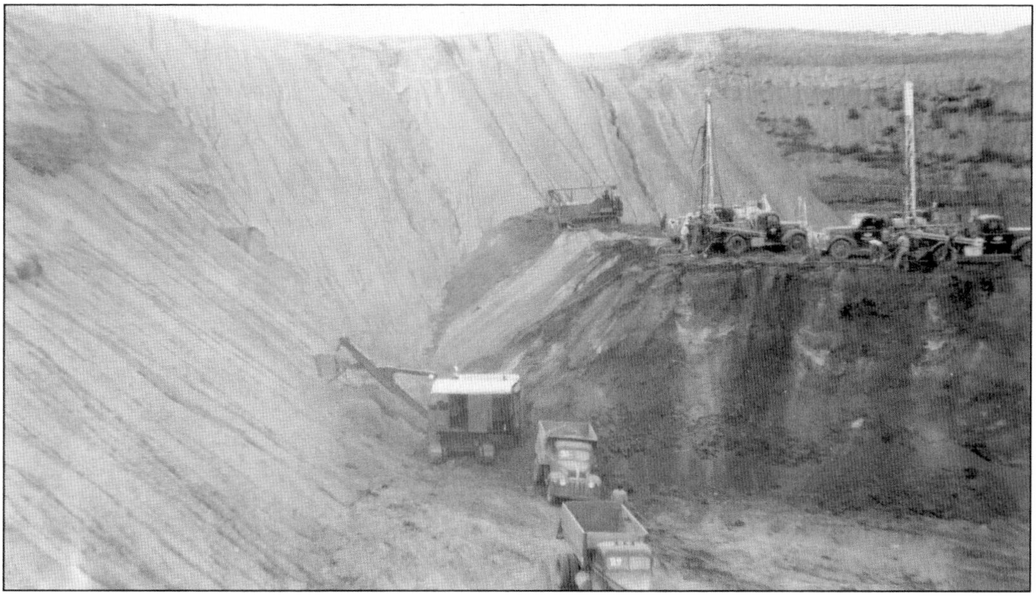

The Black Diamond underground mine became the Blue Diamond surface mine when Timber Jack Joe Lynde leased it in 1944. He leased the coal from the government for 15¢ a ton and planned to sell it to families for $1.50 a ton. Most people used five or six tons per winter, so Timber Jack Joe hoped to make a quick profit. The tall drills to the right are ready to blast more coal loose. (Courtesy of Joe Lynde.)

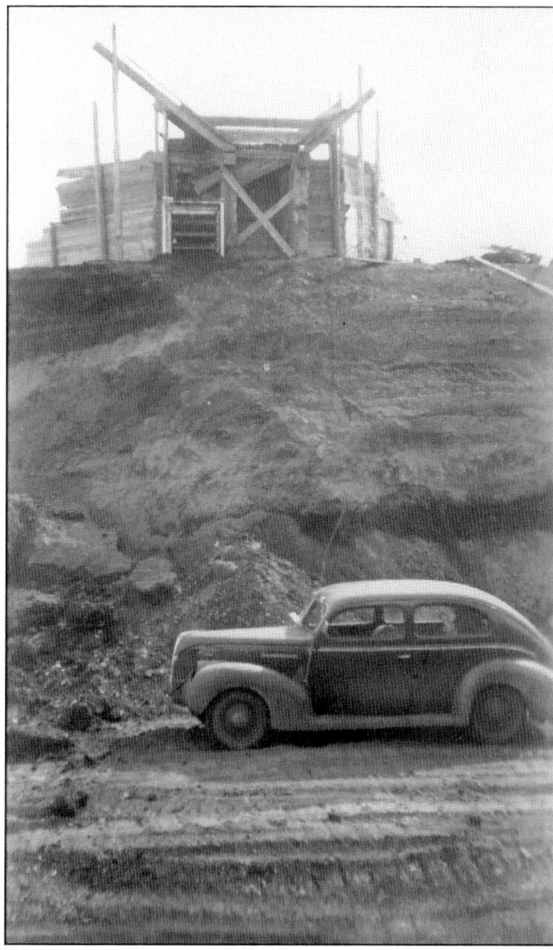

When Wyodak Mine workers were on strike in the 1940s, railroad officials approached Timber Jack Joe about picking up the coal contract, which would mean hauling 1,200 tons of coal a day. While Timber Jack Joe was in Denver finalizing the deal, a big construction company stepped in to block it. He had planned to use a railroad siding already in place under a 12-foot bank, but while he was in Denver the construction company removed the railroad tracks from the siding. One of Lynde's loadout facilities is partially built in this photograph. (Courtesy of Joe Lynde.)

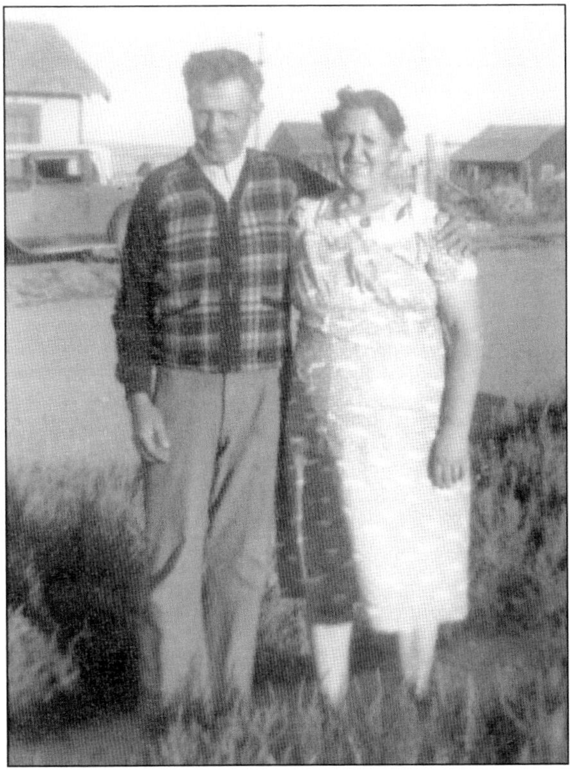

John Henry and Louisa Keck made Gillette their home in 1919. John and his brothers, Matt and Jim, operated a coal mine in which the coal was hauled from the shaft by donkey-drawn cars, then dumped in wagons to be transported away. The mine was west of Hayden Livestock, near Westover Road and 4-J. Jim Keck also worked at Peerless, where he survived a 90-foot fall and lived to be 80 years old.

Ernest E. Berry, left, was a minister, farmer, merchant, land locator, mail carrier, postmaster, county commissioner, and newspaper correspondent in the Recluse area. In December 1941, Ernest Berry and his stepson, Dorie, went to a nearby coal mine for a load of coal, and the former was caught under a fall of coal. He was badly crushed and died from his injuries. His youngest daughter, Olive, married local rancher Bill Matheny and had a son named Jack and a daughter named Darlene.

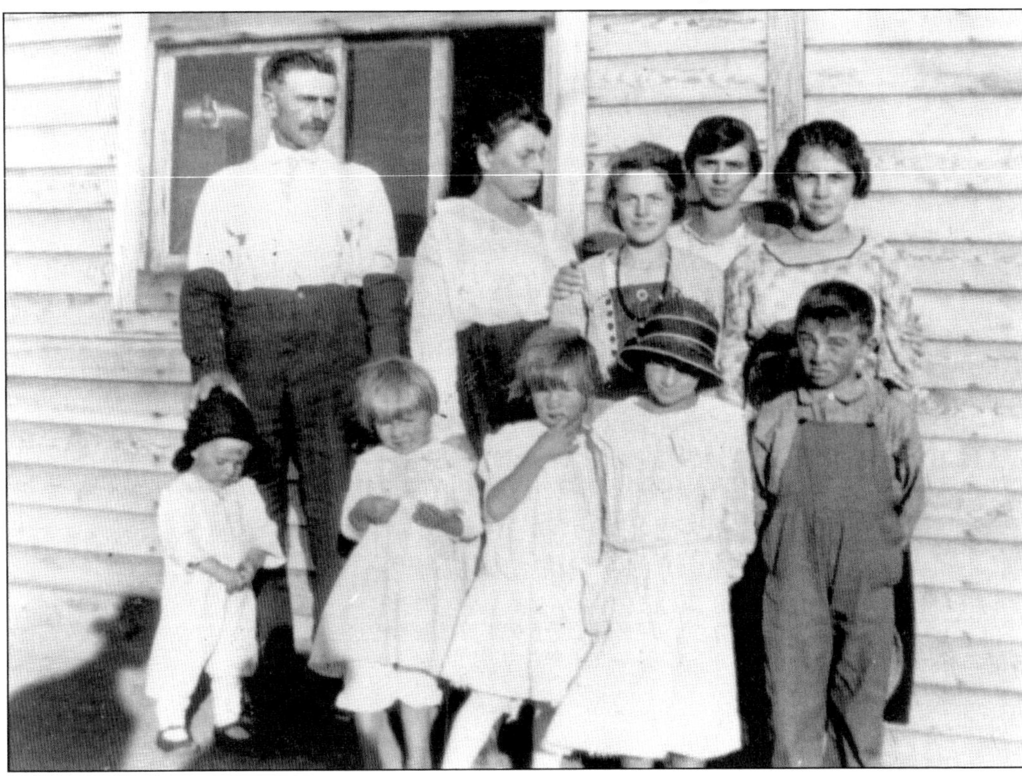

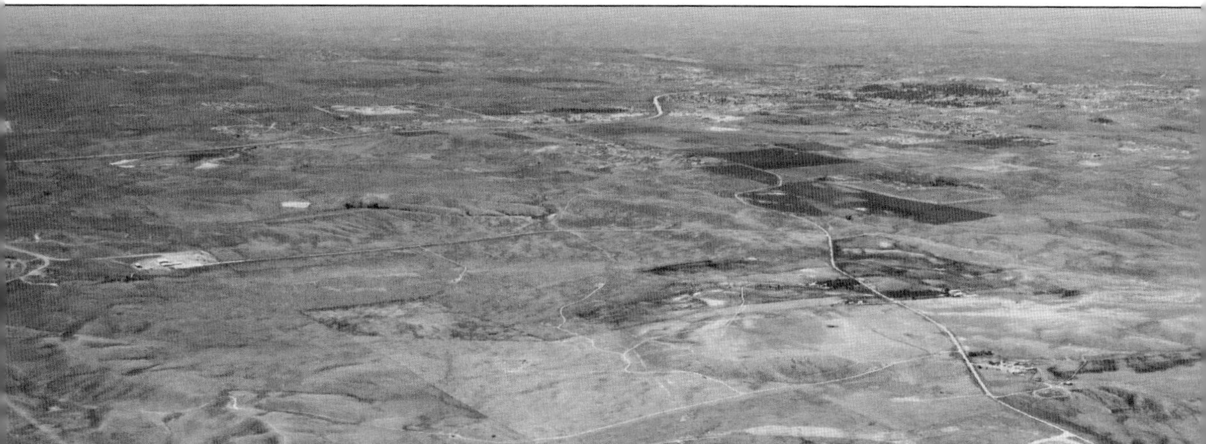

Fort Union Mine was a test mine providing coal to Denver-based Evergreen Energy, a coal technology company that uses a K-Fuel process: a combination of steam, high pressure, and high temperature to reduce the water content of less valuable coal. Powder River Basin coal has a moisture content of about 30 percent. The K-Fuel process takes that down to about 12 percent moisture. The K-Fuel process also reduces mercury and nitric oxide. The process changes the coal into high-Btu pellets that look similar to charcoal briquettes in appearance. The plant, which is close to the mine, processed about 1,000 pounds of Fort Union coal per hour, which became 500 pounds of K-Fuel. In 2010, Evergreen Energy sold the mine to Synthetic Fuels LLC's subsidiary, Dallas-based Green Bridge Holdings, Inc., but leased a portion of the plant. The mine has not produced coal since 1997, when production was approximately 440,000 tons.

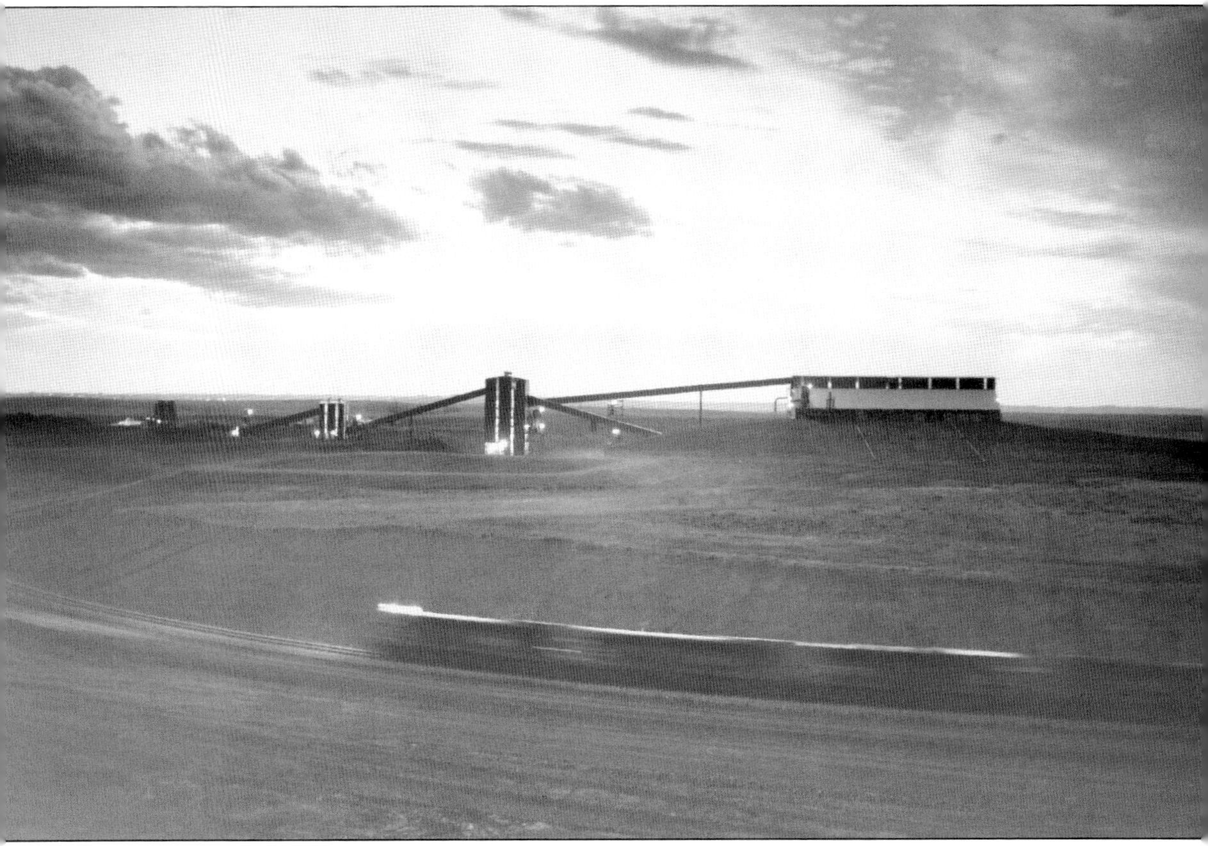

Clovis Point Mine was owned by Kerr McGee at the time shipments began August 8, 1979. All of the initial coal production was shipped to the Cajun Electric Cooperative in Louisiana. The mine was located about eight miles east of Gillette off Garner Lake Road. The mine was originally called the East Gillette Mine No. 16, but was renamed for Indian spear points called "clovis points." Three spear points, dating from the 12th century BC, were found near the mine site. The spear points, discovered in 1977 during an archaeological assessment of the area, were created by Indians first known to have lived near Clovis, New Mexico. A spear point is the mine's official symbol. In early 1987, Kerr McGee announced it would be consolidating coal operations at its Jacobs Ranch Mine, including supplying coal for any Clovis Point Mine contracts. The exposed coal at Clovis Point was covered with overburden, and all 77 employees transferred to Jacobs Ranch. The mine was "mothballed," or stabilized, so it could be reopened if economic conditions changed.

Two

The CCC Years

Between World War I and World War II, the United States went from unlimited optimism to chaos. In the 1920s, worker productivity increased more than 40 percent. The United States, untouched by battle, rapidly rose to become the world's leading manufacturing center. Farmers found an endless market for their products. Pres. Calvin Coolidge told Congress that "the country can regard the present with satisfaction and anticipate the future with optimism." But everything was not as rosy as it appeared. European countries, unable to pay off their huge debt, were not able to purchase American-made goods, but Americans produced more and more anyway. Before long, American products became too expensive for its own people to buy. Banks took on more risk and ended up failing in great numbers. Despite the postwar optimism, production slowed, putting thousands out of work. Banks failed, and mortgages were foreclosed, leaving many penniless and homeless. Pres. Herbert Hoover believed the federal government should stay out of locally solvable problems. He felt that free enterprise and private charity could support those citizens affected by the current economic setbacks. Farm foreclosures reached the tens of thousands. For those able to keep their property, the droughts and dust storms of the early 1930s wiped out any crop they were able to grow. Franklin Delano Roosevelt, then governor of New York, gathered the resources of his office to provide any possible relief to thousands of New Yorkers devastated by the economy. Among his solutions was a program to put men to work in forestry, erosion control, and improving recreational facilities. The Democratic candidate, Roosevelt, was put in office on March 4, 1933, as a result of proposing that the federal government take a prominent role in righting the nation's economy. Within days of his inauguration, Roosevelt called a special session of Congress and presented his proposal to create government agencies and programs designed to stabilize the nation's economy. On March 31, 1933, Roosevelt signed a bill into law creating the Civilian Conservation Corps, which was intended to "build men, [and] to accomplish constructive results in our vast federal, state, private forests, and other resource properties."

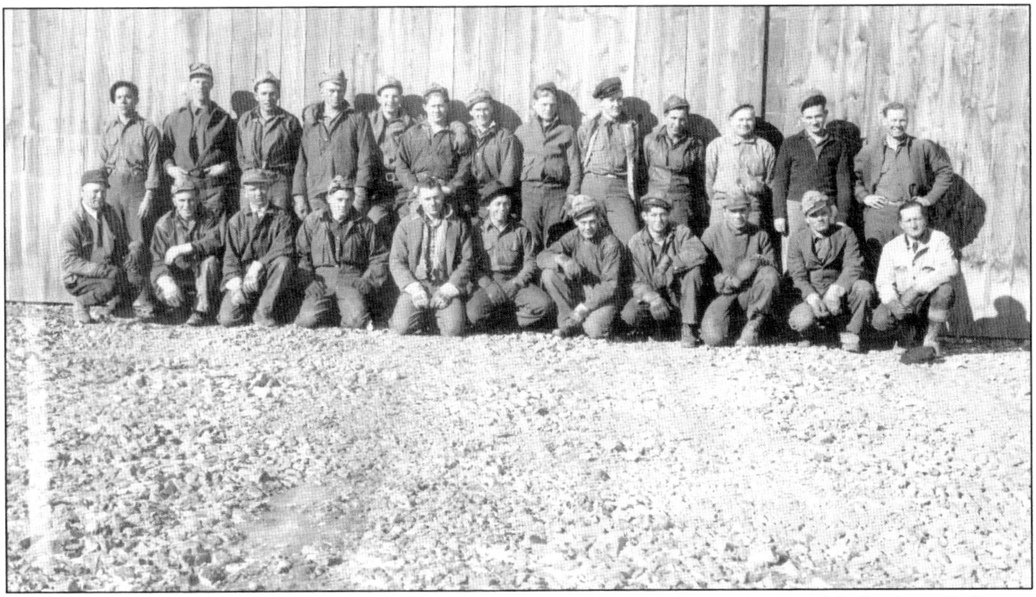

In 1933, Camp 886, the only CCC camp to fight coal-mine fires, was established in Gillette, Wyoming. With the exception of 23 local Wyoming boys, all of those enrolled at the camp were from Texas. Classes were taught at the camp, including first aid, English, mathematics, photography, Spanish, auto mechanics, and carpentry. As over half the men in the company were Mexican, it was fortunate that one of the teachers, Roy Guiterrez, was qualified to teach in Spanish.

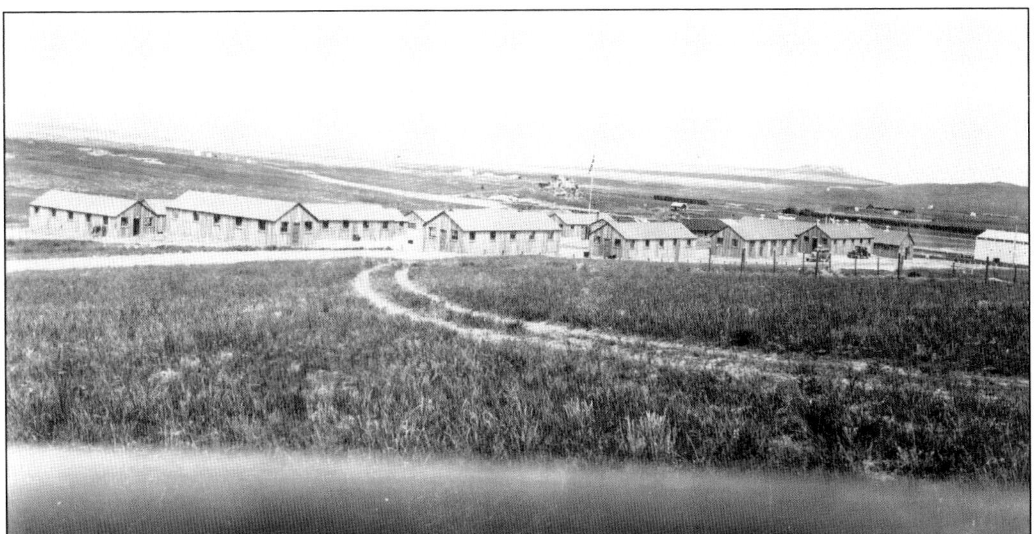

The camp in Gillette started as a city of tents, intended for summer use only. The camp was dismantled in October and set up again in April. It did not take long to build more permanent barracks to withstand the Wyoming winds and weather. When the men worked at a coal fire, the coal they excavated during the process was hauled to the camp to heat the barracks and fuel the kitchens.

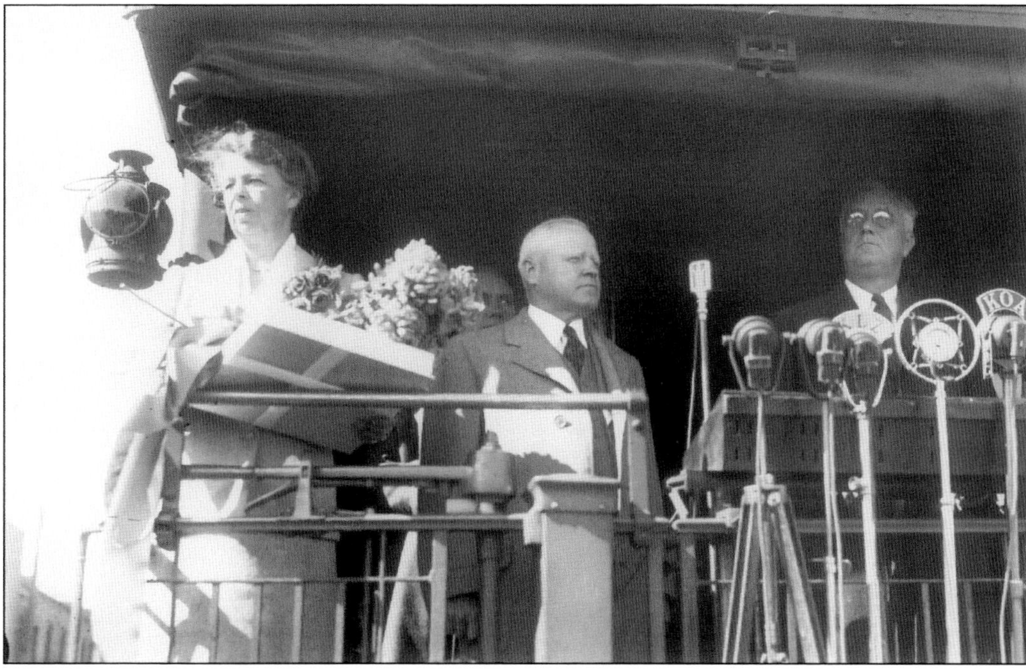

Camp 886 was named Camp Miller in honor of Wyoming governor Leslie A. Miller. Pictured in the center with President and First Lady Roosevelt, Governor Miller visited the camp in August 1933. He inspected the camp and visited the Elmore and Ditto Mines, where work of extinguishing fires was underway. The governor was a guest of the Gillette Lions Club for lunch at Camp Miller, and dinner at the Sage Café. (Courtesy of Wyoming State Archives, Department of State Parks and Cultural Resources.)

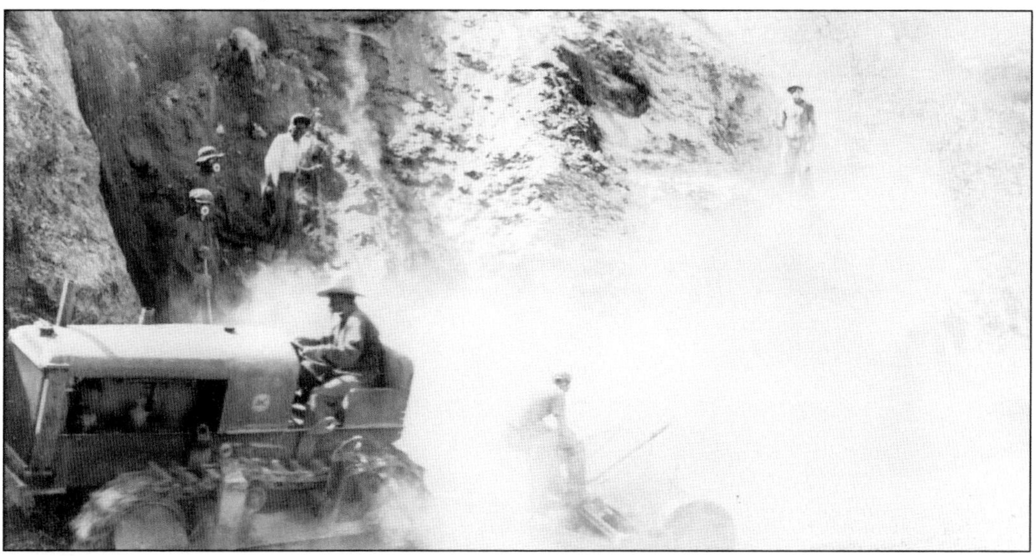

Safety and employee health are more prominent issues these days than they were in the past; however, safety around burning coal seams was a concern even during the CCC years. Not only did burning coal seams waste a precious resource, but of greater importance was the danger of children, adults, and animals falling through the weakened ground. Coal burned to a very fine ash, which required the men to wear respirators and goggles in order to function.

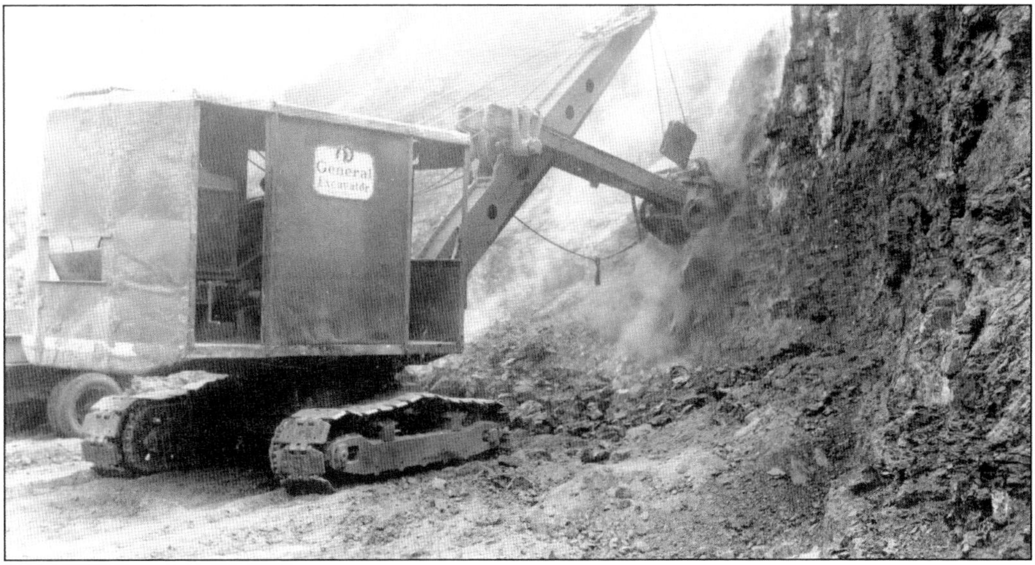

Three methods of firefighting were used. The direct method, used on small fires, consisted of removing the hot and burning material to a safe distance, mixing it with dirt, and covering the location of the fire with a layer of earth to prevent the fire from breaking out again.

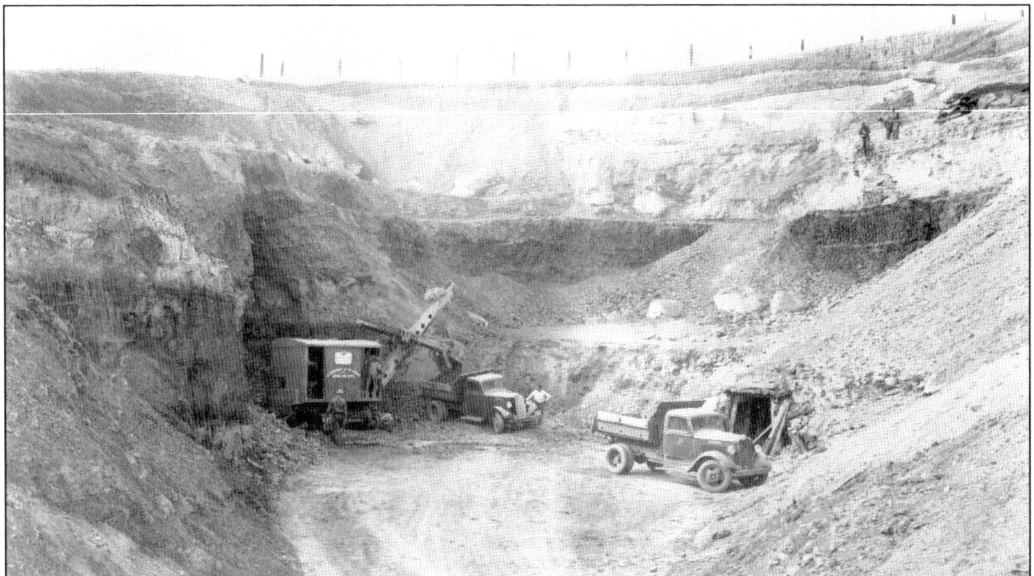

The indirect method was used when the burning vein was too deep and covered too large an area to remove the fire. Where a large vein had burned out, the ground would cave in, leaving fissures and giving air to any remaining fire. All openings were filled and tightly packed with loose dirt, causing the fire to burn out. This work could be done by hand or by steam shovel.

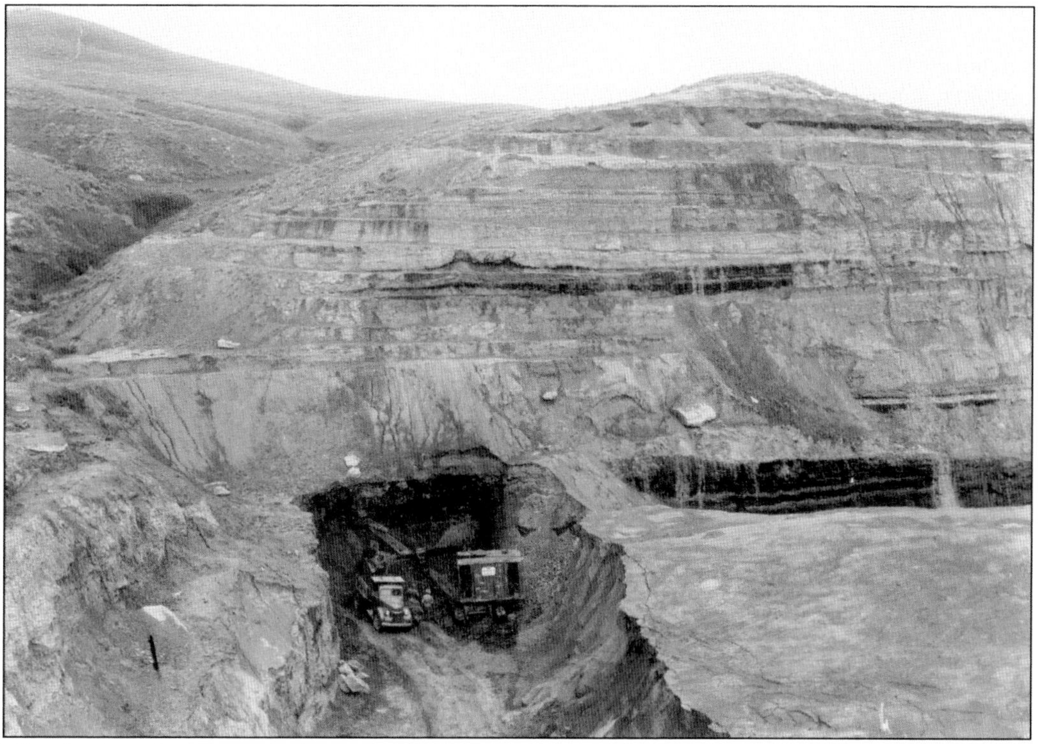

The third method was trenching, used where the extent of the burning vein is easily determined. Trenches are dug around the edges of the area, and these in turn are filled with dirt, confining the fire to a restricted section, which will, in time, burn completely out.

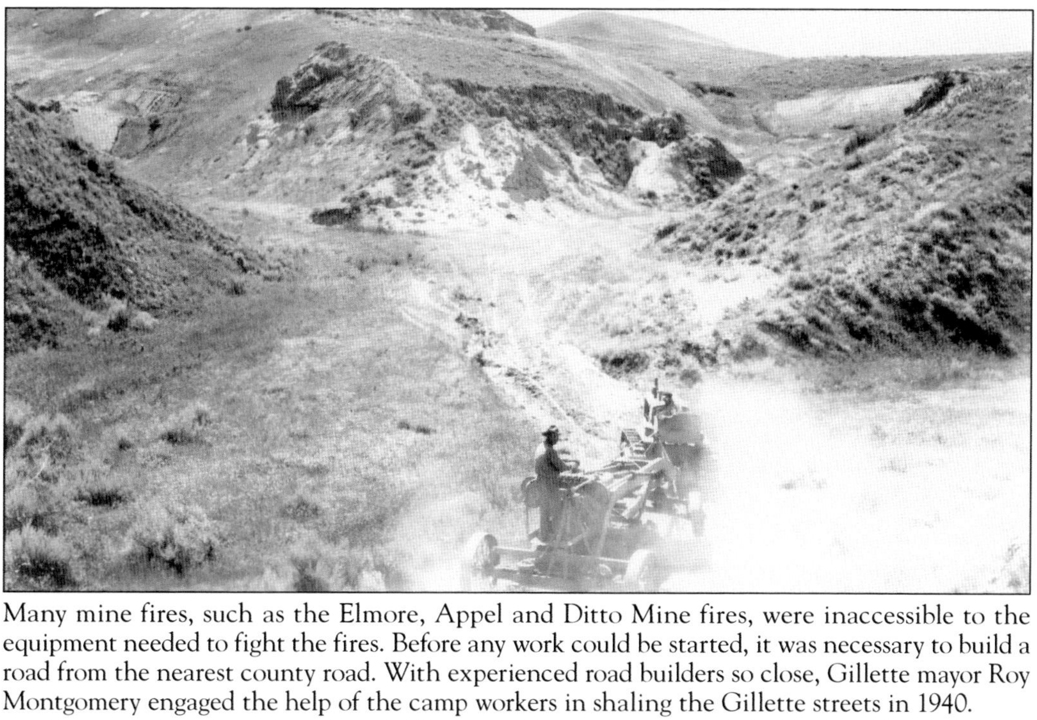

Many mine fires, such as the Elmore, Appel and Ditto Mine fires, were inaccessible to the equipment needed to fight the fires. Before any work could be started, it was necessary to build a road from the nearest county road. With experienced road builders so close, Gillette mayor Roy Montgomery engaged the help of the camp workers in shaling the Gillette streets in 1940.

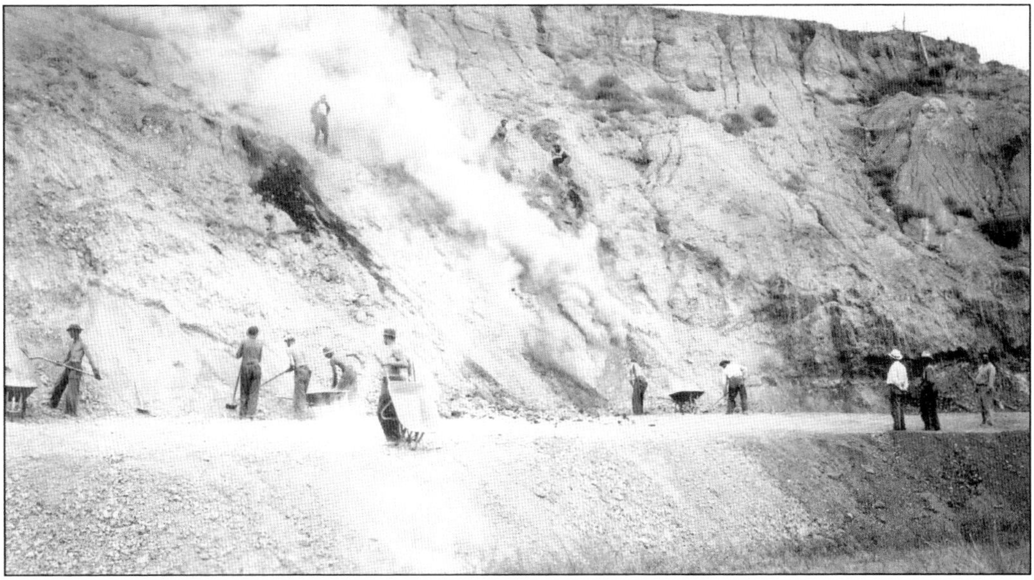

The Consolidated Report for 1933 to 1938, recapping all of Camp 886's projects, refers to some coal seam fires as "cancer" type. These were fires that had a habit of branching out in all directions. It was not unusual to treat a fire that had five or more cancer fires. Cancer fires were peculiar, in that one spot would be hot and then both the coal and the overlying dirt would be cool, only to connect with another similar spot in which the burning was active.

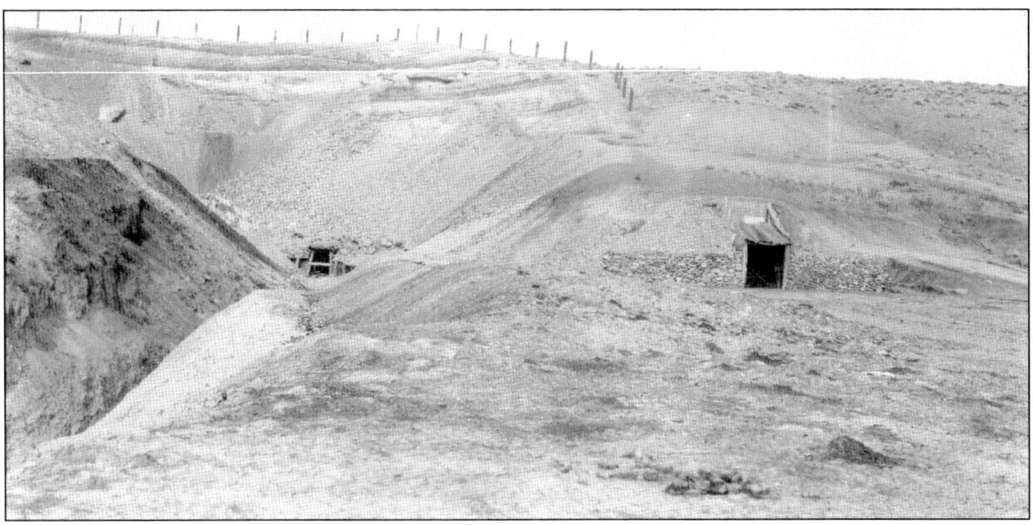

Early settlers tapped into the free and abundant coal. As settlers had to dig deeper to access the coal, portals were built, and the mine became an underground mine. Some portals were shored up with rocks and lumber; some were not. CCC workers caved in underground mines and covered portals, helping to put out the fire and to protect the residents of the area from entering a dangerous area.

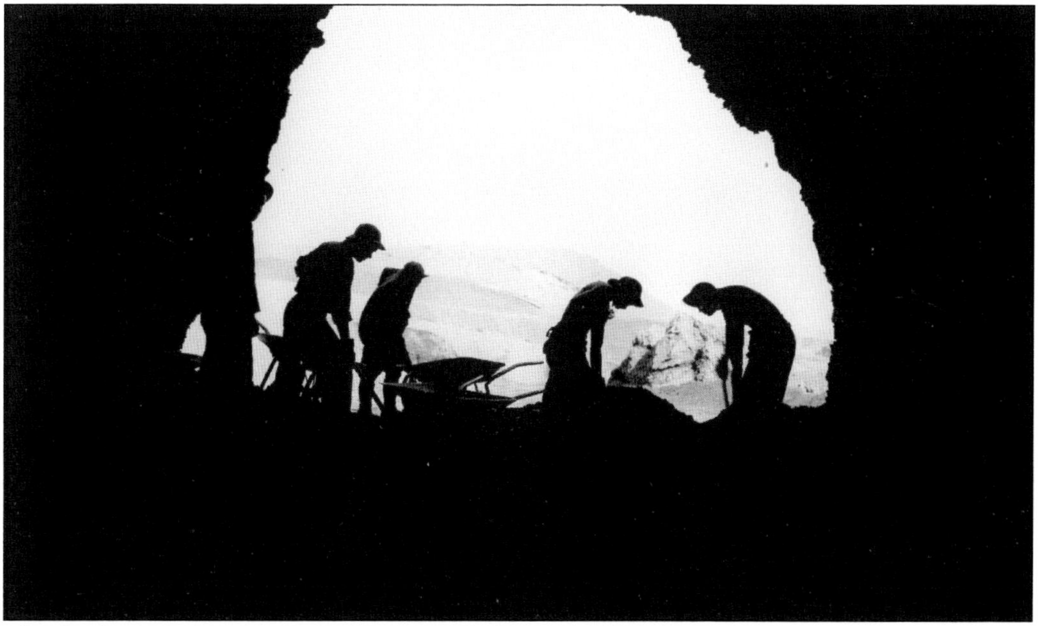

The Civilian Conservation Corps in Campbell County, Wyoming, was the only camp of its kind developed to put out underground coal fires, many of which had been burning for years. Some fires called for using large steam shovels, but in this mine opening, hand tools and wheelbarrows made more sense. The prairie homesteaders learned that where there are coal fires, there is more coal that can be used as precious fuel.

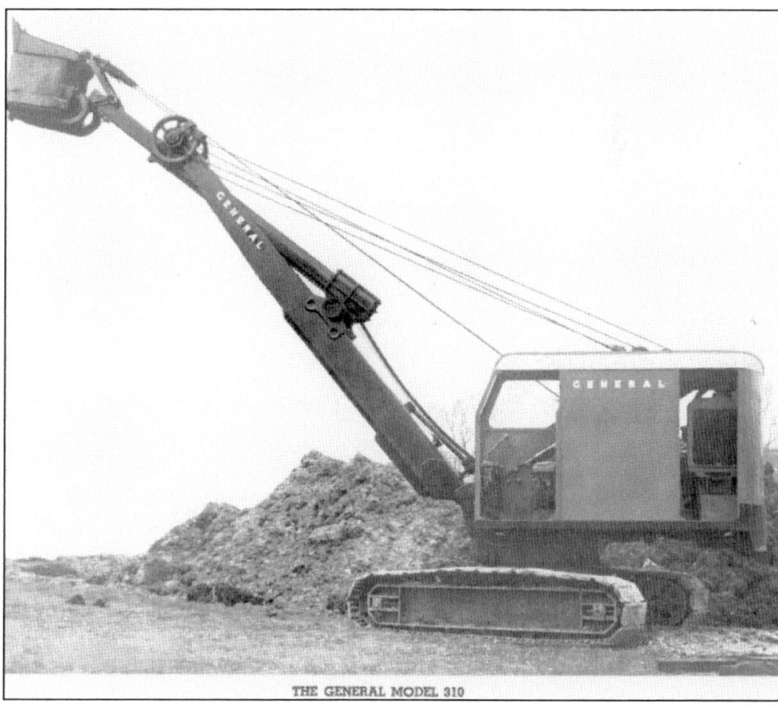

Crawler tracks were first developed in England for use on tanks in World War I. They were initially applied to a Marion shovel in 1916. The Marion Steam Shovel Company, founded in Marion, Ohio, in August 1884, produced wrecking cranes, dredges, log loaders, and railroad ditchers, in addition to shovels. (Courtesy of Marion County Historical Society, Marion, Ohio.)

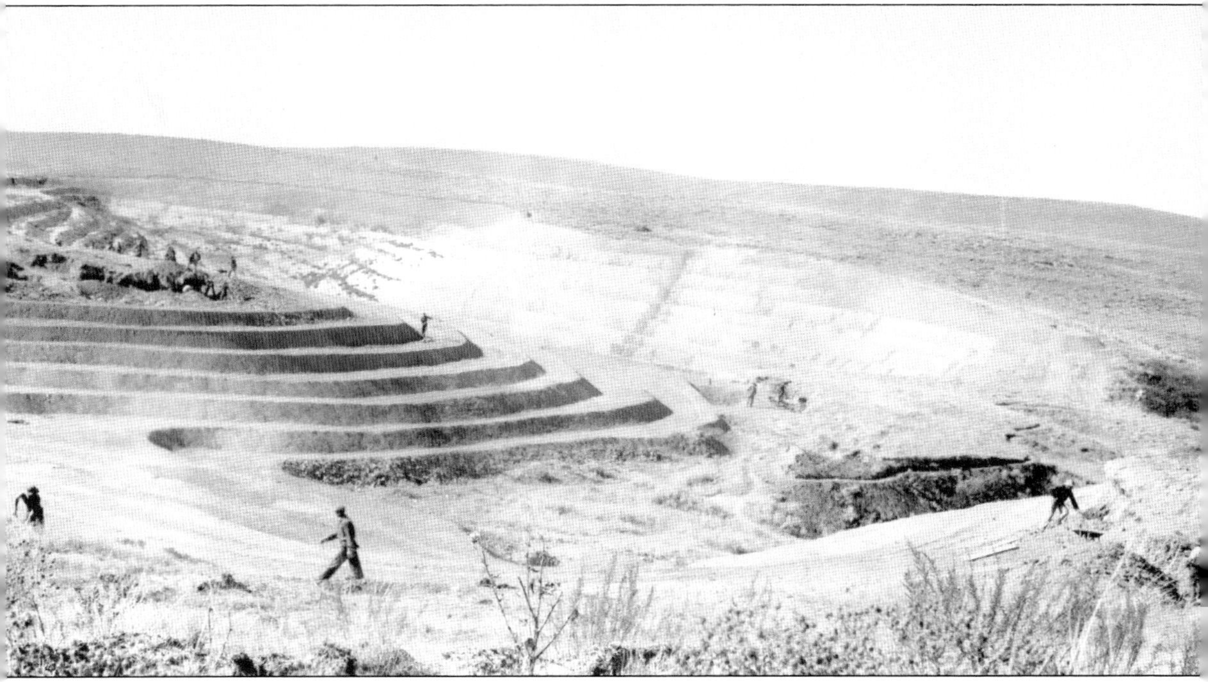

After a coal fire was extinguished, the CCC workers might have built retaining dams and terrace slopes to help with erosion control. On many of these projects, work was done by hand. The shovel was mainly used to dig out burning coal. If there was burning coal when trenches were formed, it would be covered with a thin layer of fine dirt. Mud was also applied to areas the shovel could not reach, to keep air from feeding the fire. A project like the Ditto fire required 6,223 man-shifts with 100,645 cubic yards of dirt, coal, and ash moved. Four hundred tons of coal were hauled from the location of the fire to the CCC camp. Workers learned lifelong trades—such as power shovel operating, truck driving, road grading, and firefighting—which they could take home with them after their CCC service. Millions of tons of coal were saved by the efforts of Camp Miller's residents.

Three

Ayrshire to AMAX to Alpha

Many people do not know the early history of Alpha Natural Resources. One of the oldest names is Ayrshire Collieries Corporation, which dates back to 1923. With Pierre Goodrich as chairman of the board of Ayrshire Collieries Corporation in the 1940s, the company experienced monumental growth in both coal production and sales. From the 1940s to the 1960s, profits increased, as well as the number of employees. By the time Goodrich decided to sell Ayrshire in 1968, Ayrshire was the 11th-largest producer of bituminous coal in the country. Ayrshire was sold to American Metal Climax in October 1969. At the time, American Metal Climax (AMAX) owned no coal mines; its primary business operations involved mining molybdenum, lead, and copper in the upper Midwest. Although the sale was purported to be valued in excess of $100 million, the company's true worth may have been substantially underestimated because of their inability to foresee the demand for low-sulfur coal, the growing hostility toward nuclear power, and the Arab oil embargo of 1973. One of the best things Goodrich did in his years at Ayrshire was to have Ayrshire obtain as many coal reserves as possible, including 50,000 acres in Wyoming. With AMAX pumping millions of dollars into its new investment, AMAX became the third-largest coal company in the United States by 1975. There was some doubt about whether the western mines would amount to much, but AMAX's Belle Ayr and Eagle Butte Mines carried AMAX economically through the late 1970s and early 1980s. The 1970s were a decade of huge growth with little concern for cost control or operating efficiency. That approach did not work into the 1980s, when a dozen competitive mines were developed in the Powder River Basin. The 1990s saw two changes in ownership: one in 1993 and one in 1999. On November 13, 1993, parts of AMAX, Inc., merged with Cyprus Minerals Company to form Cyprus AMAX Minerals. A subsidiary of that company was Cyprus Amax Coal Company, and its subsidiary Amax Coal West, Inc. On June 30, 1999, Cyprus Amax Coal was bought by RAG International. Foundation bought RAG in 2007; and Alpha Natural Resources bought Foundation on July 31, 2009.

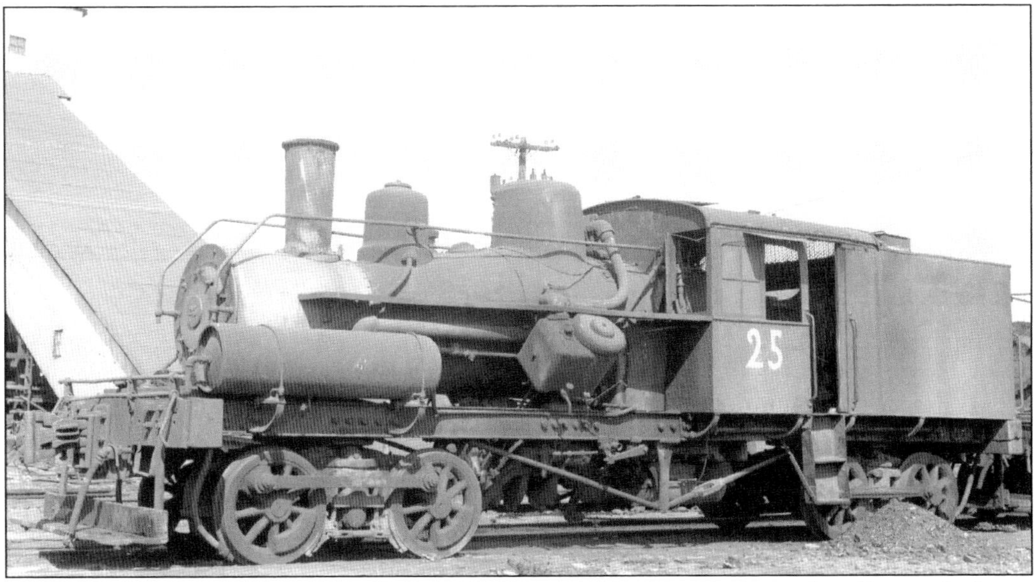

The coal mines that made up Ayrshire Collieries Corporation were located in Illinois, Indiana, Kentucky, and Wyoming. When Pierre Goodrich began the process of selling Ayrshire Collieries, two leading contenders were Ashland Oil and Refining Company of Ashland, Kentucky and American Metal Climax (AMAX) of Greenwich, Connecticut. When a deal with Ashland fell through, AMAX immediately stepped up to buy it. Ayrshire Collieries locomotive No. 25 was part of the early days of Ayrshire Collieries.

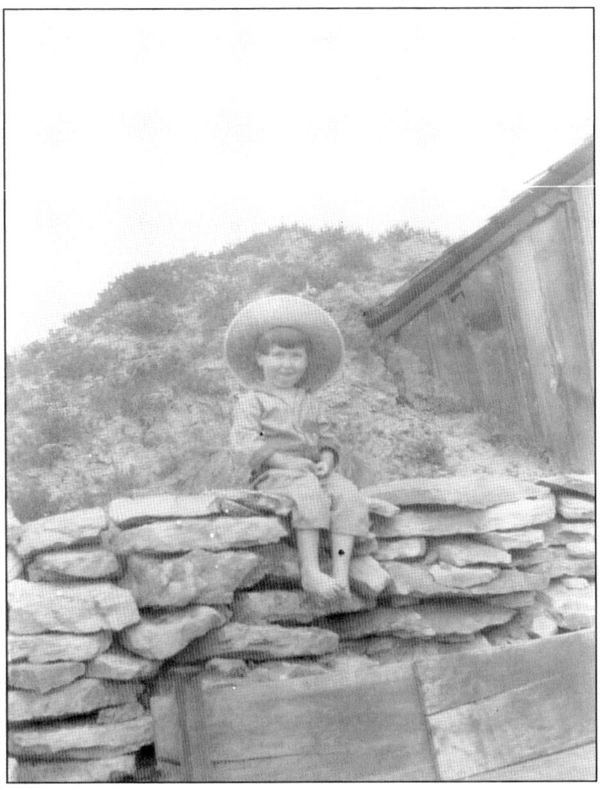

Mary Fulkerson sits on a stone wall at her parents' homestead, just north of Gillette, around 1932. The hill behind her was called Eagle's Nest. What a coincidence that 44 years later, Mary's cousin Pam Fulkerson won the contest to name the new AMAX coal mine north of the airport. The name she suggested was Eagle Butte, for which she won a $100 savings bond. (Courtesy of Fulkerson-Keck photograph collection.)

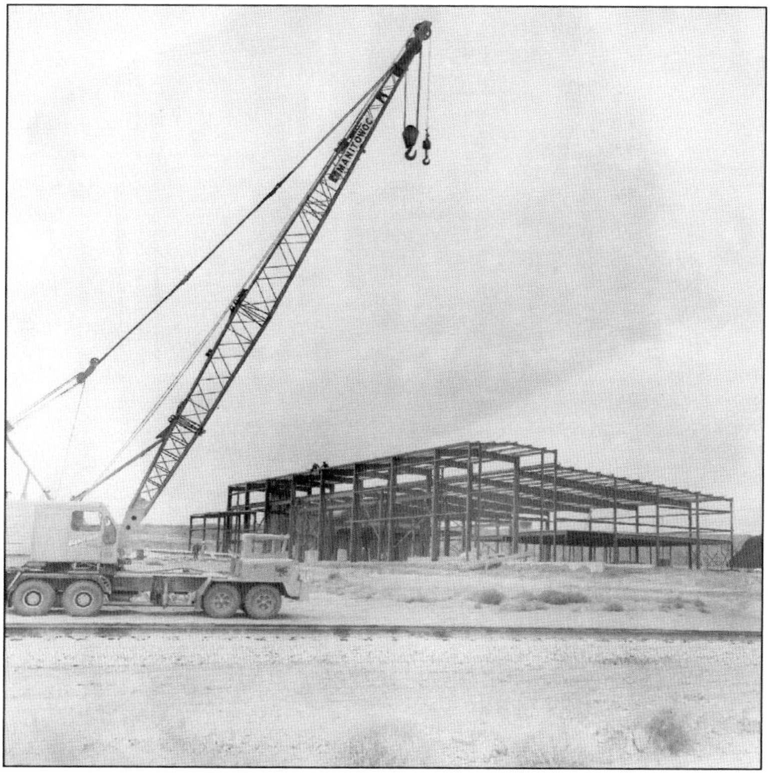

Belle Ayr Mine was located on the Lester Clabaugh ranch, about 18 miles south of Gillette. Ayrshire Collieries, headquartered in Indianapolis and Evansville, Indiana, acquired the original Federal Coal Leases in 1965 for what would become Belle Ayr and Eagle Butte Mines. The administration building shown here was constructed in 1972 at a cost of $450,000. The name Belle Ayr is derived from the Belle Fourche River and Ayrshire Collieries.

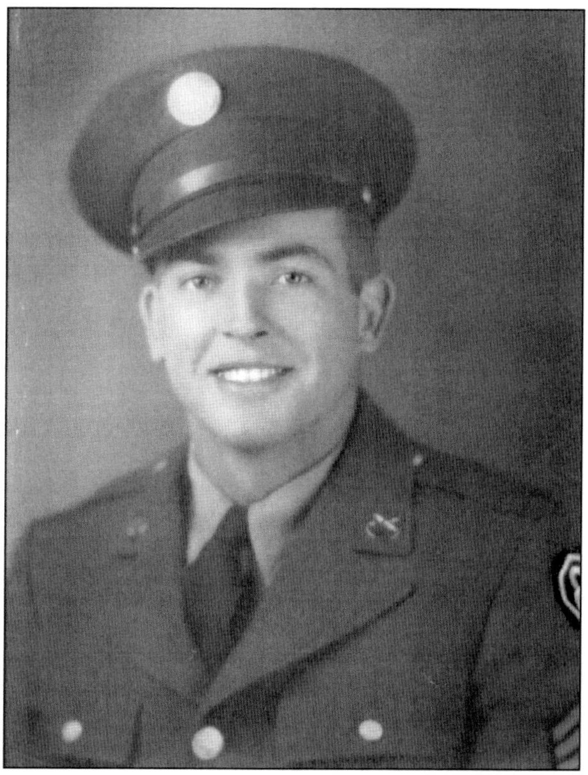

Walter Hollie Hopper was instrumental in the exploration and development of Powder River Basin coal. His start in the mining industry was shoveling and weighing coal in Illinois for 40¢ an hour. During the Depression years, Hopper went to a Civilian Conservation Corps camp in Necedah, Wisconsin, where he was paid $45 a month. In 1941, he joined Ayrshire, staying there until the Army called him to World War II. (Courtesy of Michael and Jody Hopper.)

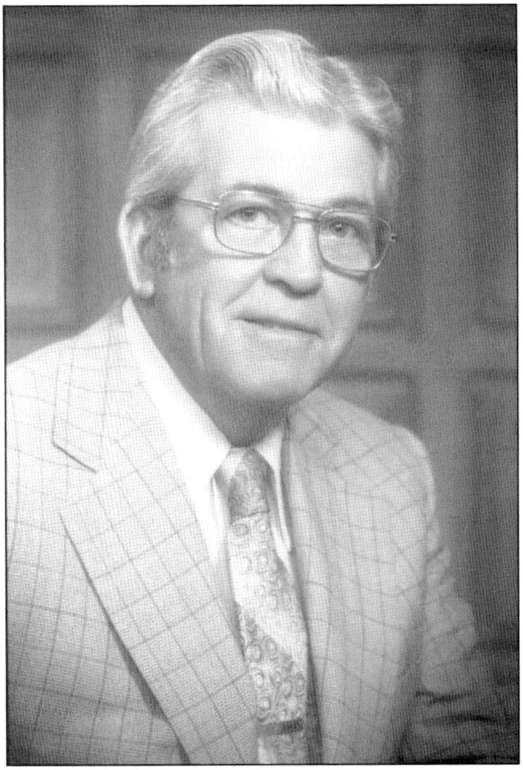

Ayrshire's merger with American Metal Climax or AMAX put the company in the enviable position of having the capital to modernize and add equipment, develop its management, offer new coal for new contracts, and to continue its commitment to reclamation, health, and safety. Hopper, shown in both images on this page, worked out of the Indianapolis-based AMAX Coal Company headquarters. (Courtesy of Michael and Jody Hopper.)

On March 21, 1975, Pres. Gerald R. Ford met with representatives of coal-related industries in the White House. President Ford stressed three objectives during his presidency, comprising the following: to increase America's capability to convert coal into clean gas and liquid fuels; to develop technologies to burn coal without any environmental damage; and to improve mining technology so we can produce coal safely and environmentally responsibly. (Courtesy of Michael and Jody Hopper.)

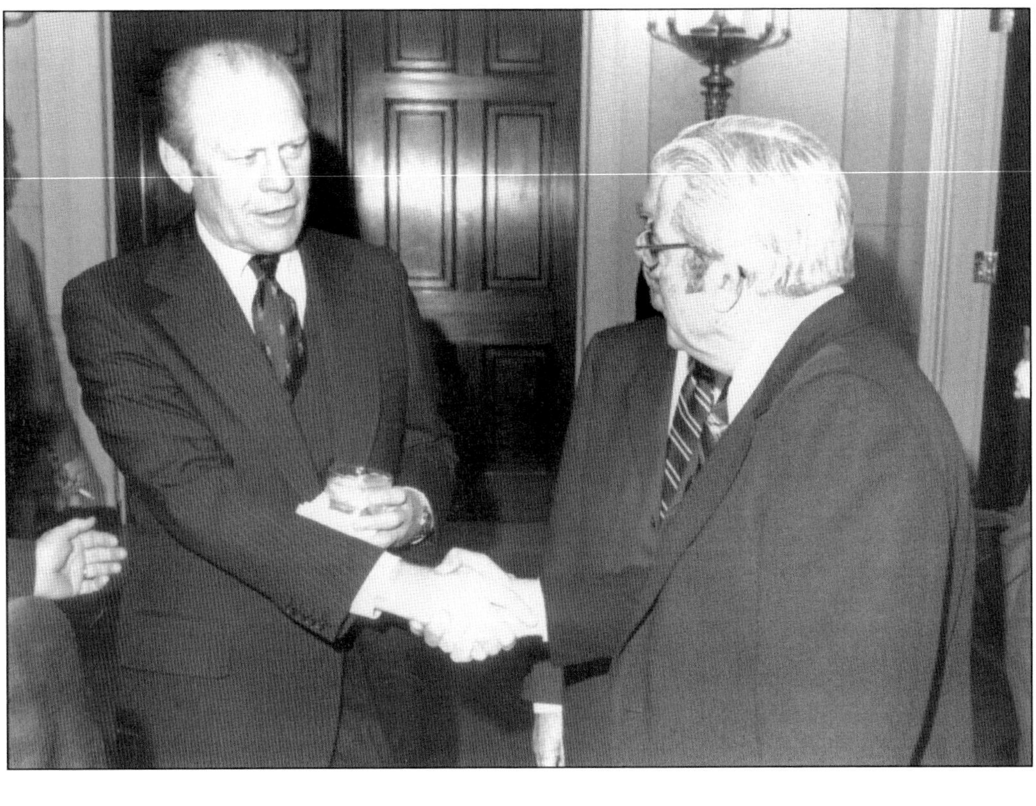

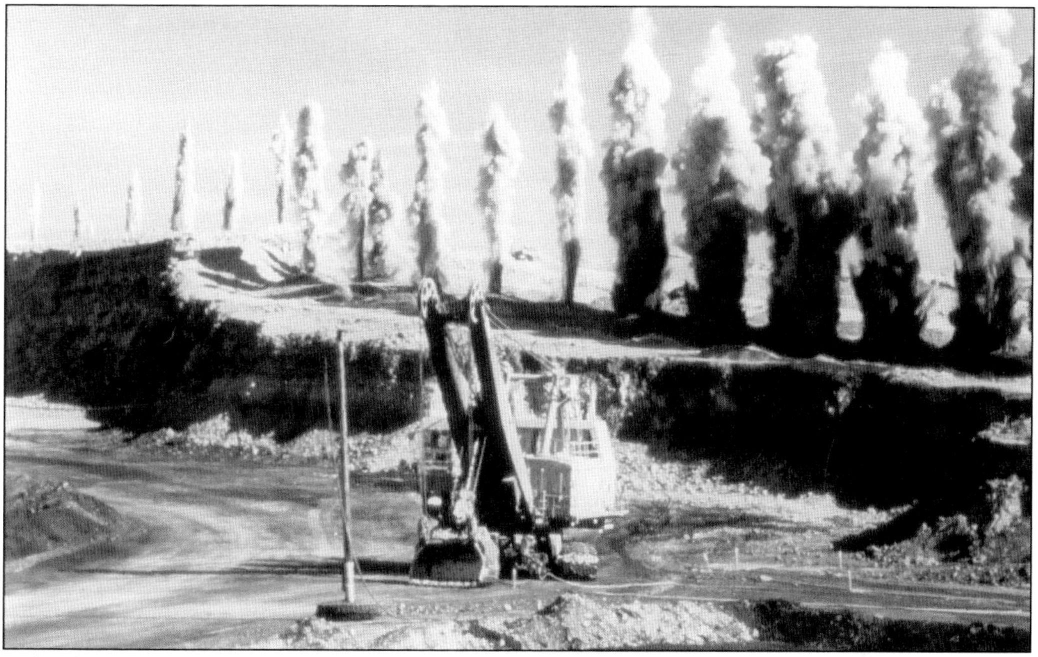

"Chuck's 1st blast," written on the back of this photograph, refers to Charles Sumner, who has been in blasting at Belle Ayr for over 30 years. Construction of Belle Ayr started in 1970, and the first coal was shipped to Iowa Power and Light as a test burn on October 27, 1972. Coal contracts allowed shipments of coal from either Belle Ayr or Eagle Butte Mines. (Courtesy of Alpha Coal West, Inc.)

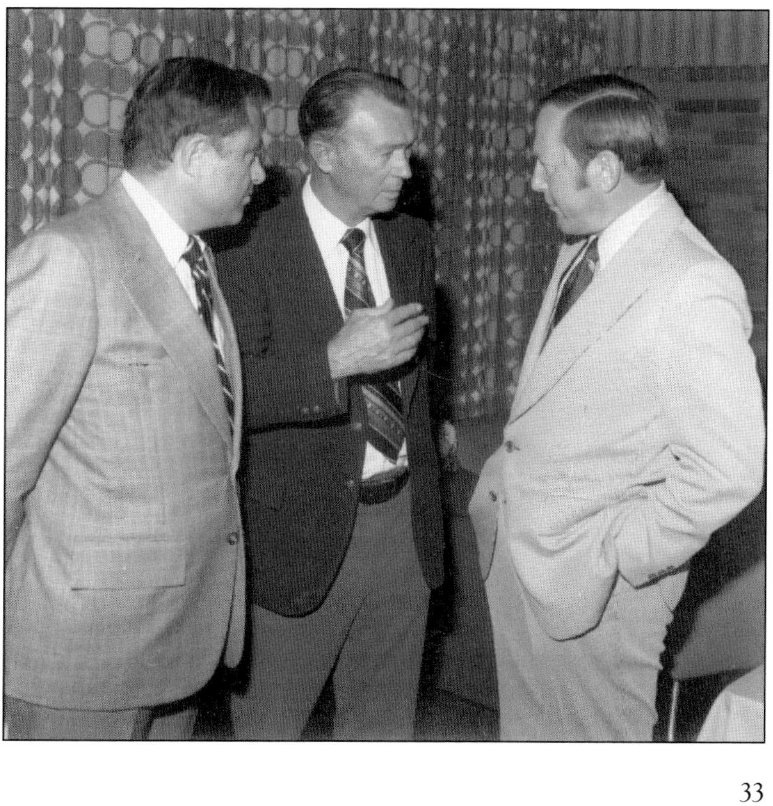

Bob James, center, was the first mine manager at Belle Ayr Mine, which started as a United Mine Workers of America (UMWA) mine. In 1974, the UMWA won an election to have representation at the mine, but in 1975 the union was defeated after a long strike. An overwhelming 95 percent of the employees voted for non-union, and the mine has been non-union ever since. (Courtesy of Alpha Coal West, Inc.)

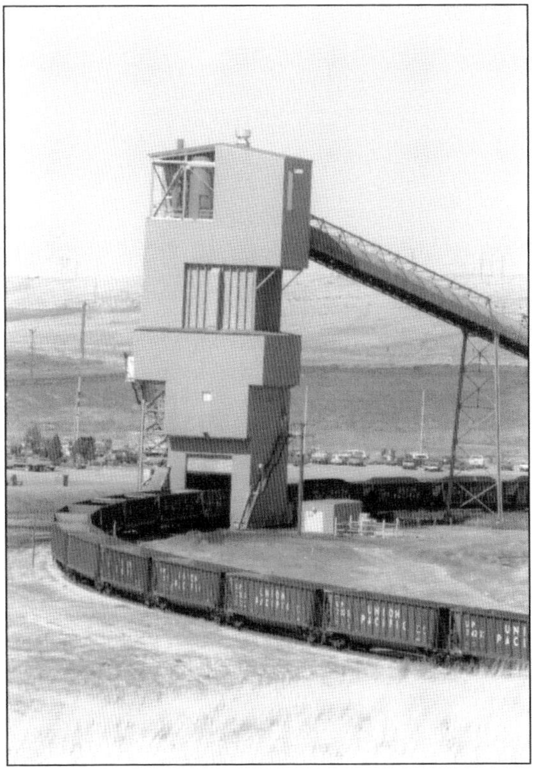

Trains travel approximately nine-tenths of a mile during loadout. There is a pacesetter on the train that acts like cruise control on a car. A batch loadout facility, like the one shown here at Belle Ayr from 1987 to 1988, weighs each train car, and loads the correct amount of coal in it automatically. A side release chemical is sprayed in the car to help the coal come out in cold weather. (Courtesy of Alpha Coal West, Inc.)

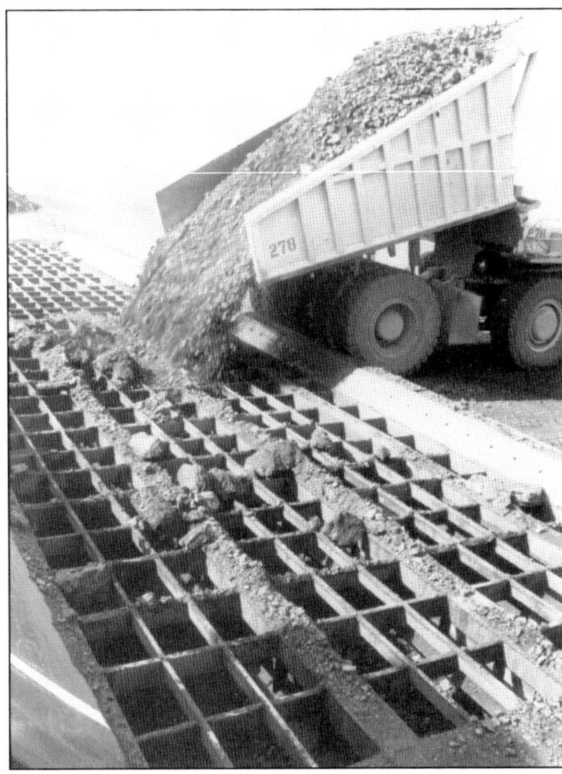

Haul trucks dump into a "grizzly," or grate, over a vibrating feeder. Many grizzlies work as separation equipment, separating coal by size or sorting rock and other debris from the coal. Such measures result in plant crushers working more efficiently. (Courtesy of Alpha Coal West, Inc.)

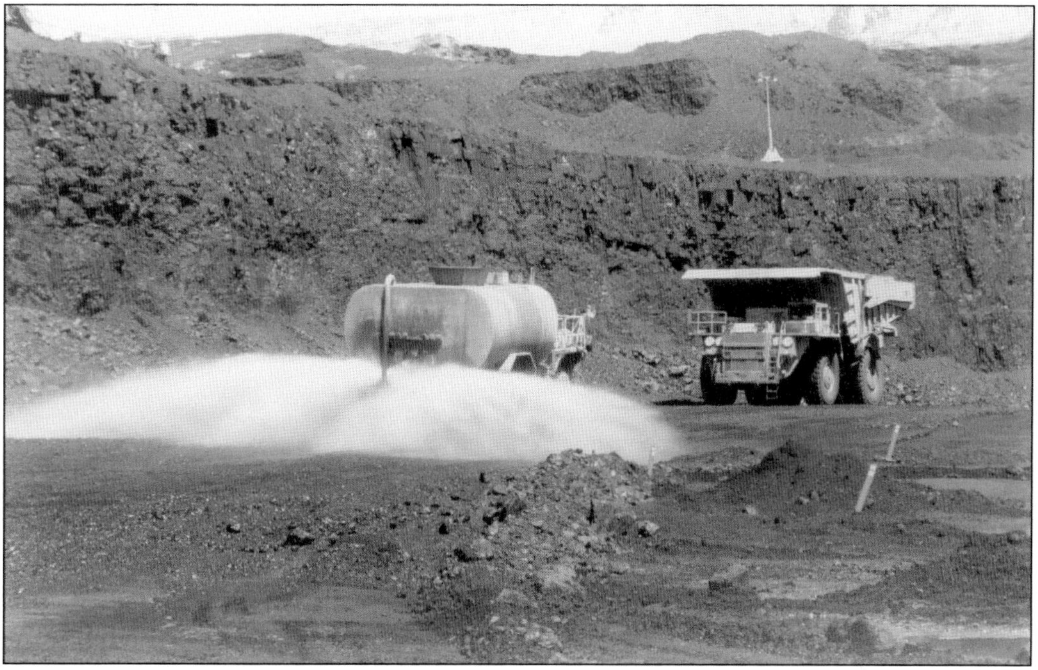

Environmental protection is extremely important at all Campbell County mines. Air quality is monitored, and roads are watered on a regular basis to suppress haul road dust. The Clean Air Act of 1990 was enacted by Congress to control air pollution on a national level. Most mining companies were practicing responsible air pollution controls before regulation required it. (Courtesy of Alpha Coal West, Inc.)

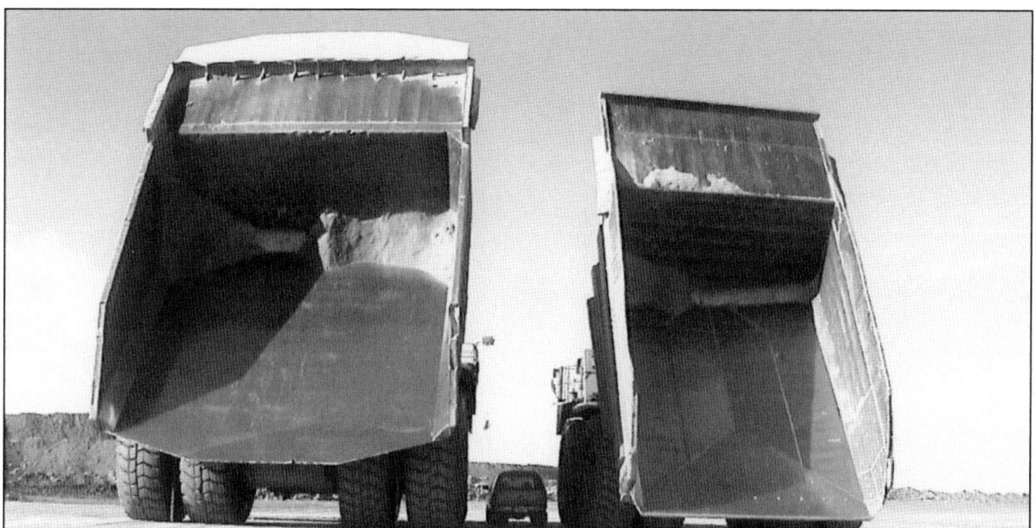

Two haul trucks, a Caterpillar and a Terex, dwarf a personal vehicle, especially when the truck beds are up. The first contract for Belle Ayr coal was with Colorado Public Service for shipments, starting in 1973. Another early shipment was 10,000 tons as a test shipment to Kansas Power & Light. (Courtesy of Alpha Coal West, Inc.)

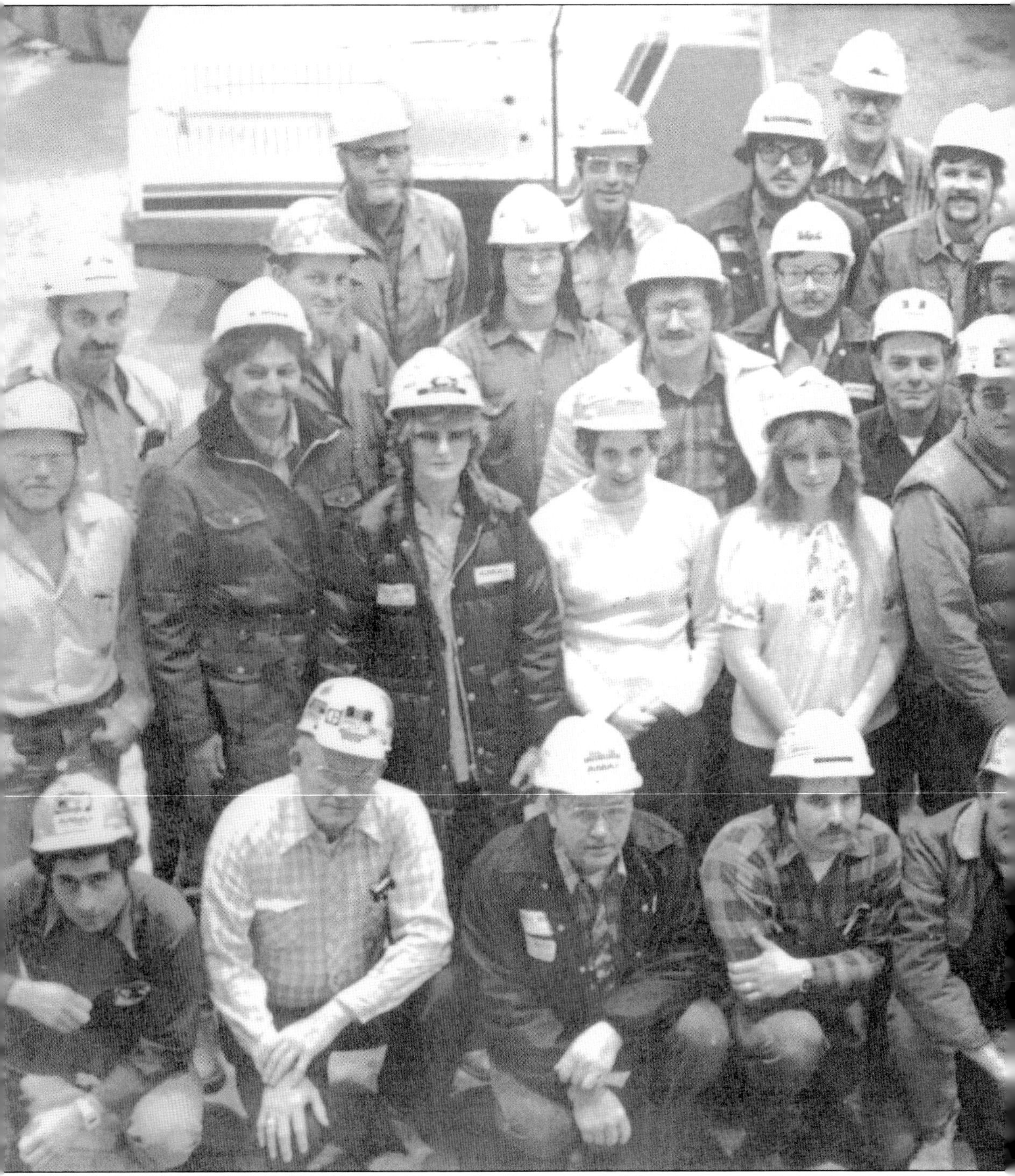

One of the first crews at Eagle Butte took a moment to pose for a photograph when Eagle Butte first opened. Pictured from left to right are (first row) Tom McGee, C. Ernie Swinney, B. "Red" McGillis, Bernie Damori, Keith Sanford, Dale O'Donnell, Willie Ponce, Ming Quan, and Jay Reynolds; (second row) Steve Kasparek, Pete Larson, Cheryl Young, Peggy Zimdars, Chris Baity, Jim Halberstadt, Charlie Madsen, Bob Jones, unidentified, and Fred Kain; (third row) unidentified,

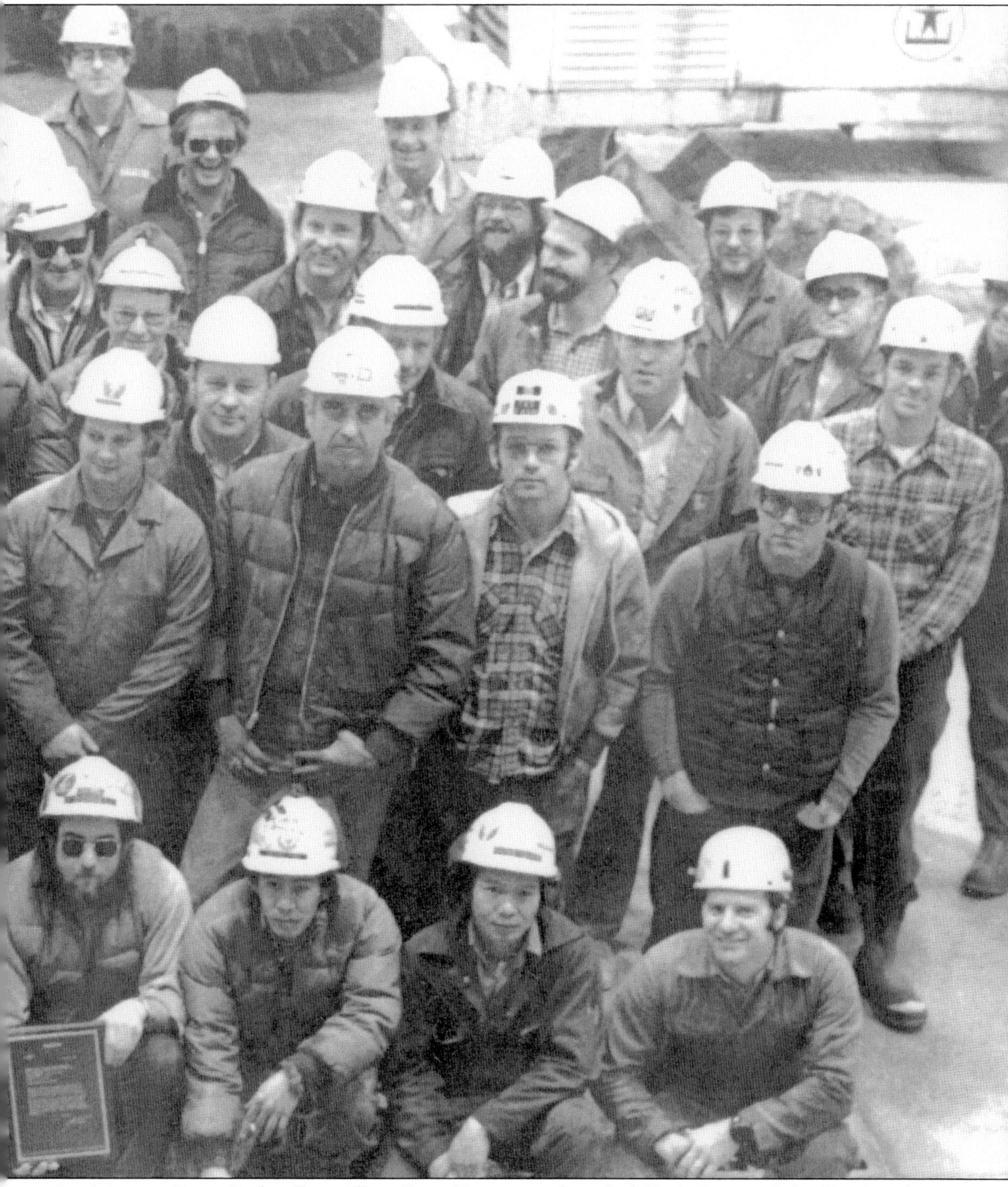

Jim Mansur, Ernie Bailey, Eric Allen, Dave Young, unidentified, Rick Tanton, John Borthwick, Paul Martin, Fritz Roberson, Don Klug, Sam Allen, and Jim Bechtel; (fourth row) Don Wood, John Reynolds, Fred Jones, Larry Miller, Jim Kotchwar, Luther Meyer, Rick Davis, John McCarty, Ron Dale, and Odis Lish; (fifth row) Dick Oelke, Arlin Martin, Ken Applegate, Craig Randall, and Mike Holcomb. (Courtesy of Alpha Coal West, Inc.)

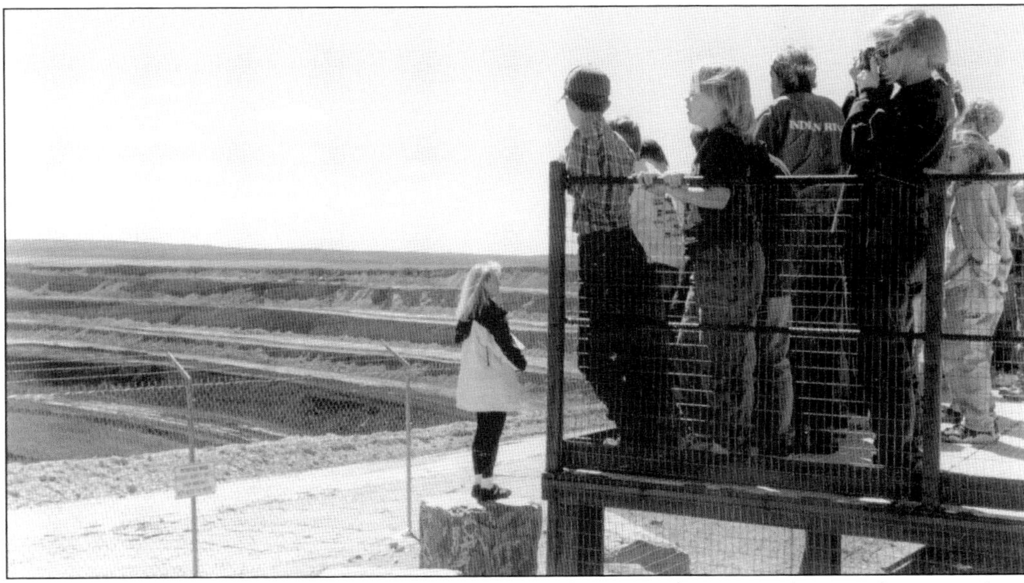

Alpha has long been a leader in public mine tours, helping to educate residents and visitors about the coal mining process, reclamation, and safety. Betsy Altenberg has guided hundreds of tours for visitors through the Campbell County Convention and Visitors Center, and for schoolchildren. For those who cannot take a guided tour, Alpha maintains a year-round overlook at the Eagle Butte Mine. (Courtesy of Alpha Coal West, Inc.)

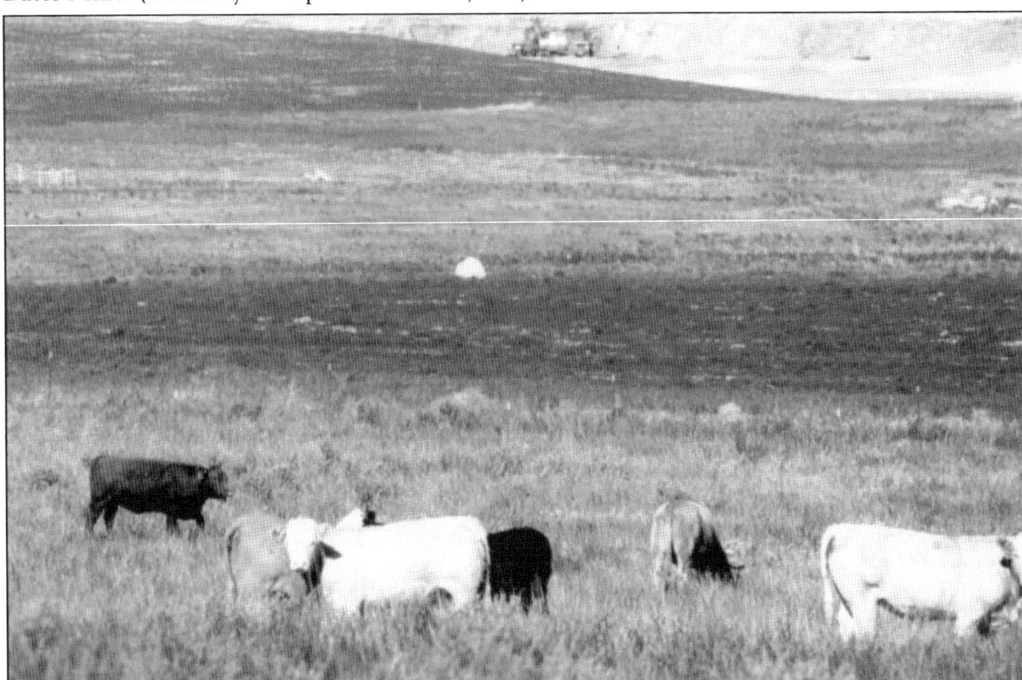

"Hollie's Hill" at Belle Ayr, named for AMAX president Hollie Hopper, was the first area mined and the first area reclaimed. New reclamation methods were developed that would work in semiarid Campbell County. Wheat was sown on land that was mined less than a year before, minimizing the time that land was out of agricultural production. (Courtesy of Alpha Coal West, Inc.)

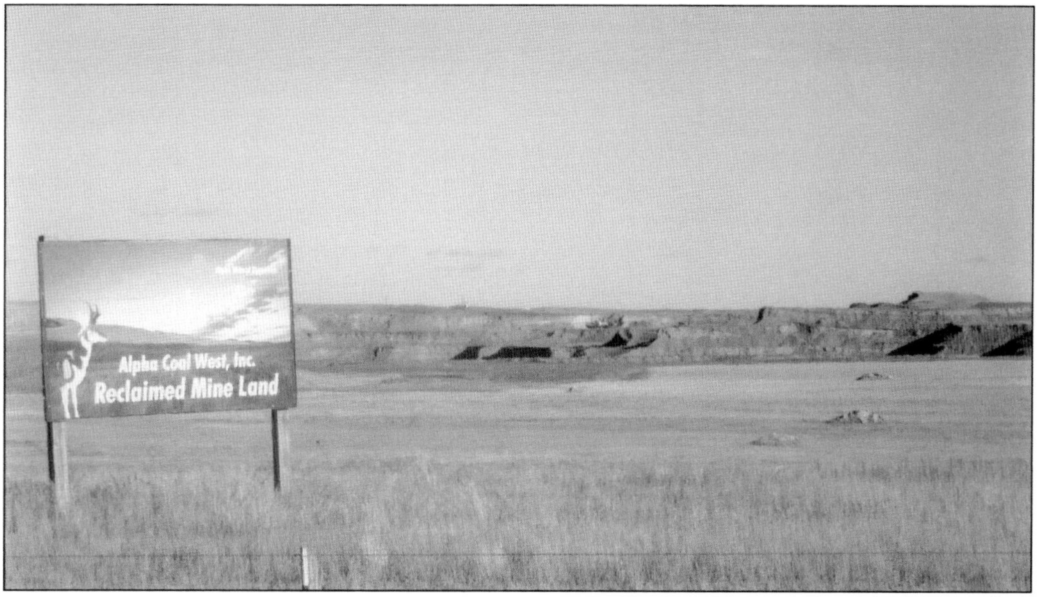

At Eagle Butte Mine, reclaimed land blends into the surrounding prairie as good as or better than it was before the coal underneath was mined. Meadowlark Farms was developed in 1945 as the environmental and reclamation arm of the company, because Pierre Goodrich, the Ayrshire chairman, was land-wise. He saw what land could do for the company as a continuing productive asset, and he was willing to invest in what was needed to learn the art of reclamation.

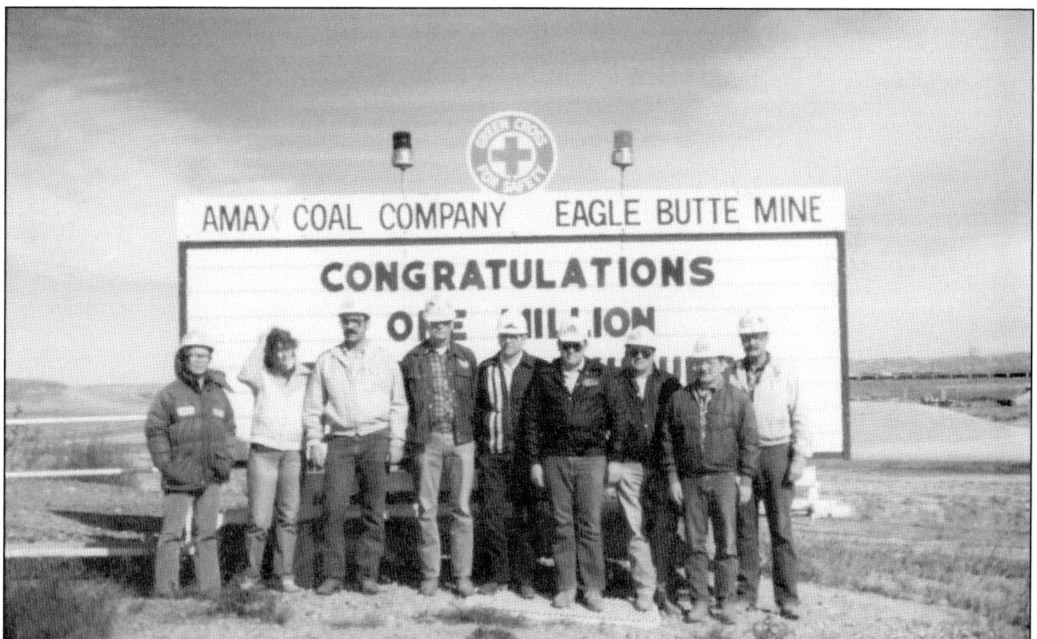

Eagle Butte celebrated one million safe man-hours on October 1, 1986. Celebrating this milestone are, from left to right, Ming Quan, Connie Baker, Mark Moser, Tom Butler, Bob Church, Clarence Bishop, Harold Pond, Chris Chastain, and Wayne Gray. The mine manager in 1986 was Tom Lien for part of the year and Bill Kirchner for the remainder of the year. (Courtesy of Tommie Butler.)

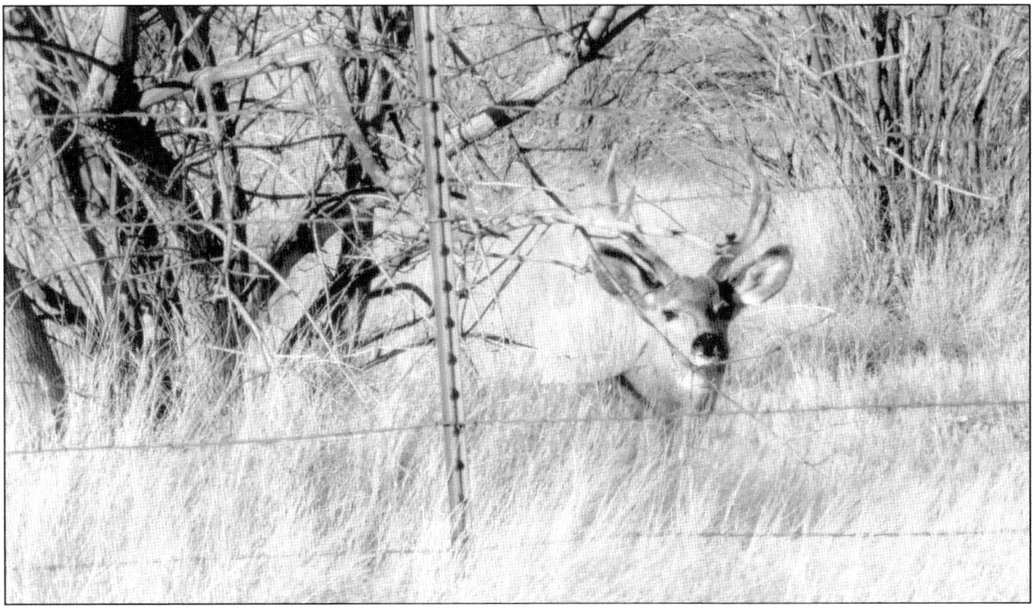

Coalbed methane is a natural gas extracted from coal beds. In recent decades, it has become an important source of energy in the United States, as well as other countries. In 1987, Rawhide Village, a neighborhood near the entrance to Eagle Butte, was introduced to coalbed methane gas before the industry knew what a valuable resource it was. Today, the deer and a few houses are all that remain of Rawhide Village.

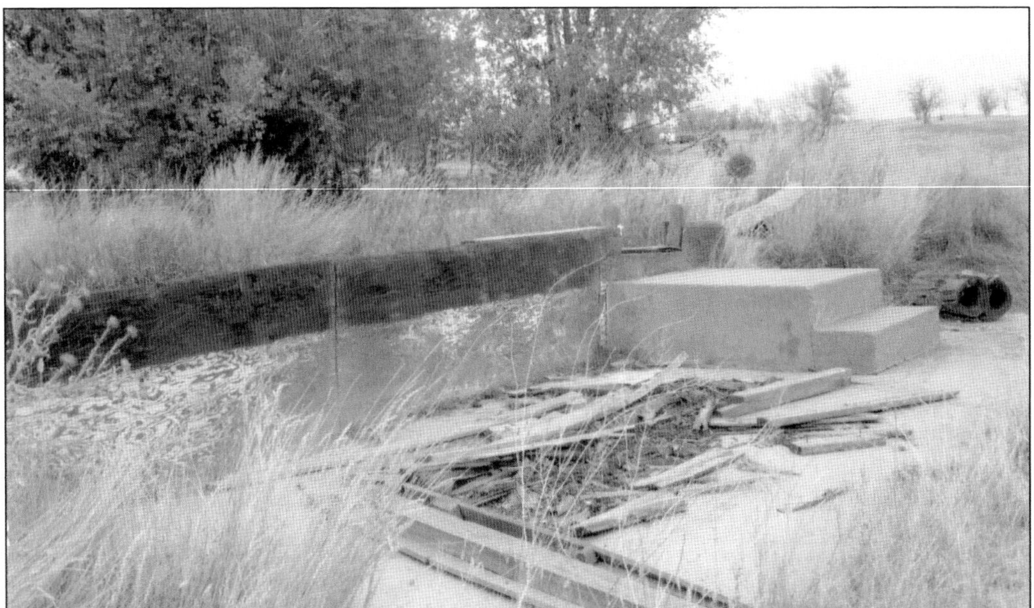

When unexplained illnesses and the distinctive rotten-egg odor of hydrogen sulfide became evident, the county commissioners ordered the evacuation of the Rawhide Village residents. Many houses were moved; some residents who felt there was little or no danger stayed put. Just a few years later, coalbed methane extraction would result in a huge boom to Campbell County's economy as more was learned about its capture and use.

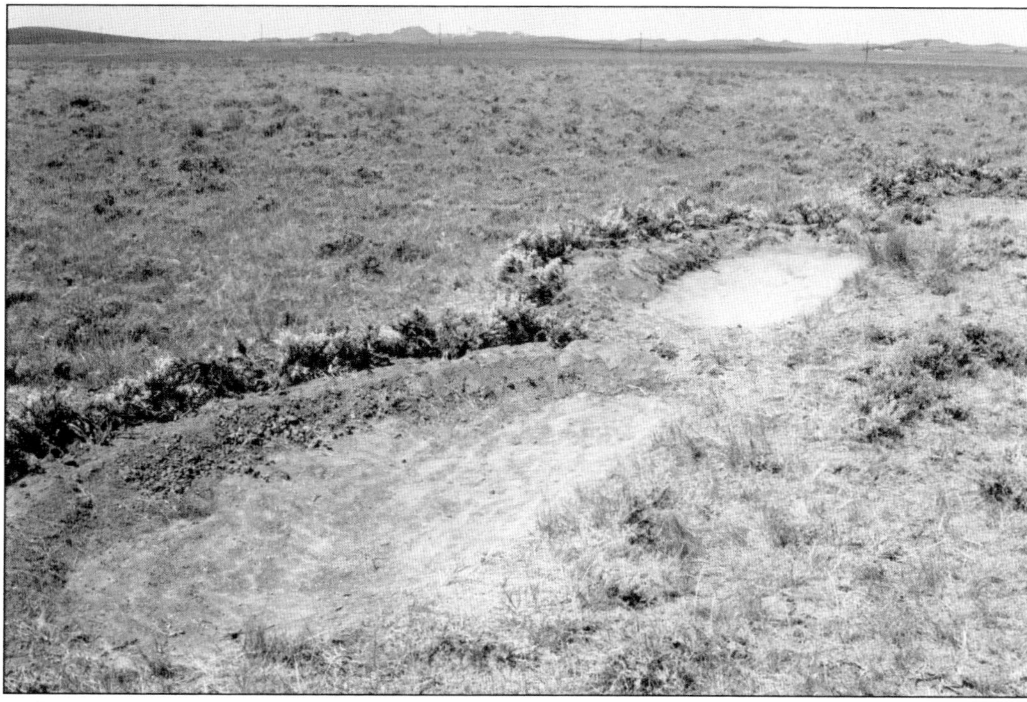

Rifle pits were found on Belle Ayr property from the Sawyers Wagon Road Expedition of 1865. Two hundred soldiers, led by Col. James A. Sawyers, were trying to find a shorter route to the gold fields of Montana. Indian attacks made the Bozeman Trail too dangerous. Historians believe the soldiers camped near Caballo Creek and built these pits as defense against Indian attack.

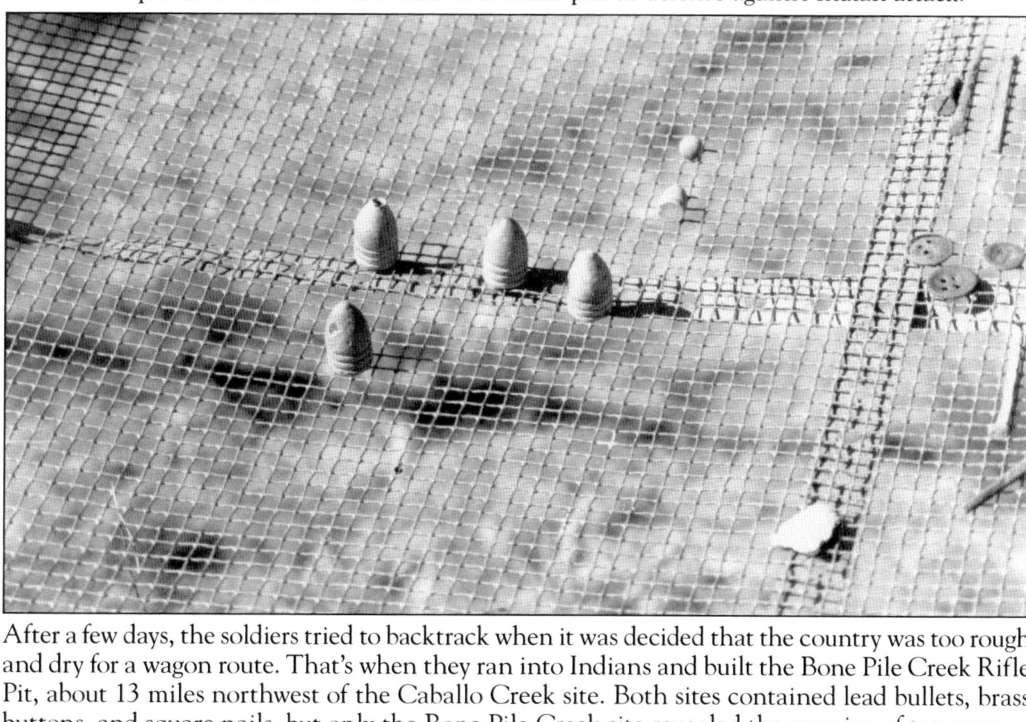

After a few days, the soldiers tried to backtrack when it was decided that the country was too rough and dry for a wagon route. That's when they ran into Indians and built the Bone Pile Creek Rifle Pit, about 13 miles northwest of the Caballo Creek site. Both sites contained lead bullets, brass buttons, and square nails, but only the Bone Pile Creek site revealed the remains of two men.

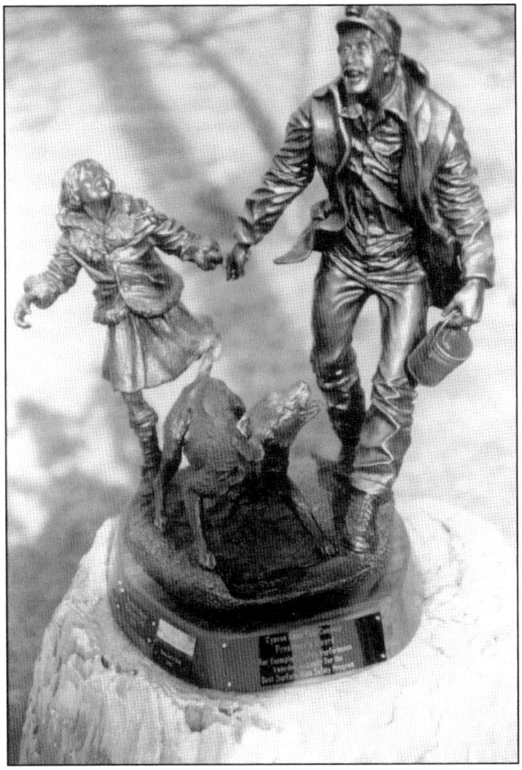

How do you say AMAX? A memo from management in 1965 stated that AMAX is a contraction of American Metal Climax. AM comes from the first two letters in American, and AX comes from the last two letters in Climax. This means that the first A in AMAX is pronounced as the letter *a* in cat. The MAX is pronounced as max in maximum. Shown here is the President's Award for Exemplary Safety Performance. (Courtesy of Alpha Coal West, Inc.)

From Ayrshire Collieries to Alpha Natural Resources, the company has consistently been a leader in community involvement. When the National High School Finals Rodeo is held in Gillette, on a regular rotation of every fifth and sixth years, Alpha Natural Resources is prominent in sponsorship of rodeo events. Alpha also commits financial support and volunteers for sporting events, human services, and quality of life events. (Courtesy of Alpha Coal West, Inc.)

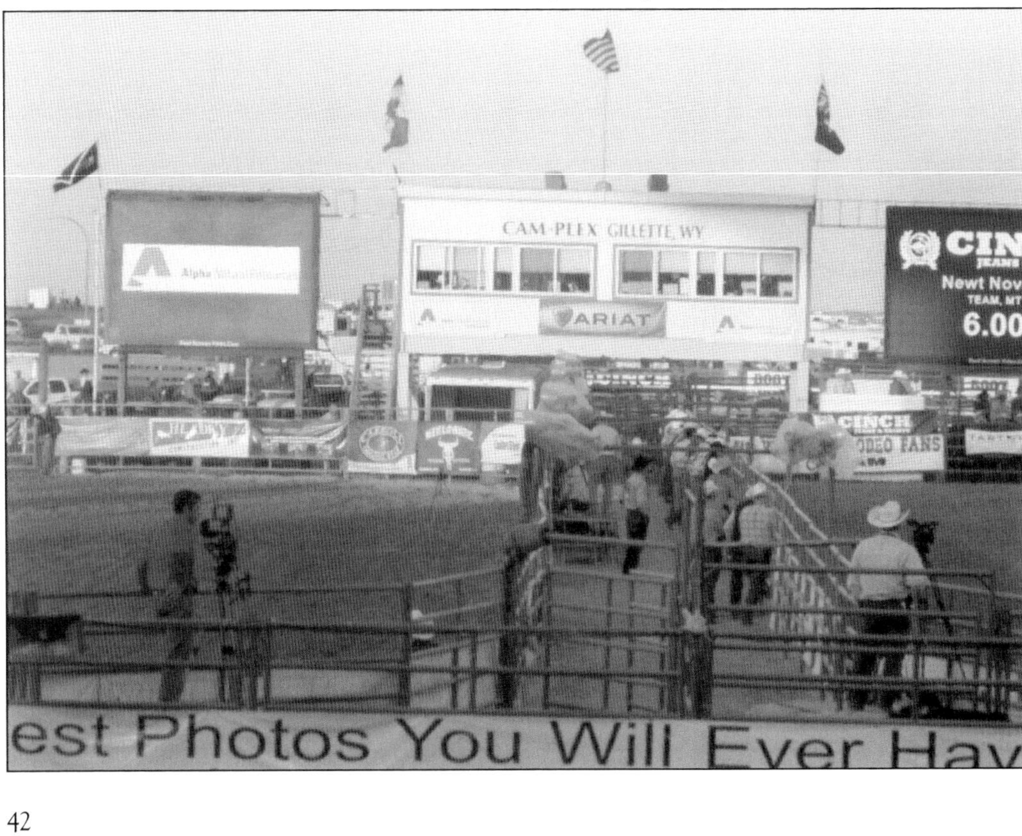

Four

Sun Oil to Cloud Peak

In 1890, the Sun Oil Company of Ohio was producing, transporting, and storing oil, as well as refining, shipping, and marketing petroleum products. Diversification of this growing company included shipbuilding and the production of oil field equipment. The mining business attracted Sun in 1941, so they formed the Cordero Mining Company in Nevada to provide mercury for Sunoco motor oils. Mercury became a vital commodity during World War II. In December 1970, Cordero Mining Company outbid nine other firms competing for coal leases on federal land near Gillette. Their bid of $3.3 million for a 6,560-acre tract was a new record for federally sold coal leases. Cordero Mine was developed in 1975 on this land, 18 miles south of Gillette. The name Cordero is a Basque word meaning "lamb," in recognition of the Basque sheep ranchers in the region. To reflect the diversification of the company, Sun Oil Company was renamed Sun Company in 1976. Cordero Mining Company was sold in June 1993 to Kennecott Coal Company. That same year, Kennecott purchased NERCO assets, which included the Antelope and Spring Creek Mines. In 1994, Kennecott Coal Company was renamed Kennecott Energy and Coal Company. In 1997, Kennecott acquired Caballo Rojo Mine from Drummond Company. Caballo Rojo Mine was Cordero's neighbor to the north. The merger of those two mines created Cordero Rojo in 1997, with combined output only exceeded by Arch Coal's Black Thunder Mine. Kennecott Energy and Coal Company acquired the 9,284-acre Jacobs Ranch Mine from Kerr McGee Coal Corporation in 1998; that property was later sold to Arch Coal interests. In 2006, Kennecott Energy and Coal Company changed its name to Rio Tinto Energy America to reflect its affiliation with Rio Tinto. In 2009, Rio Tinto spun the most significant of its Powder River Basin coal assets off into Cloud Peak Energy, a company in which Rio Tinto briefly held a 48.3 percent stake. Cloud Peak went public in 2009, becoming the only company on the New York Stock Exchange that is headquartered in Wyoming. Cloud Peak owns and operates another property—Spring Creek Mine in the Powder River Basin in Montana.

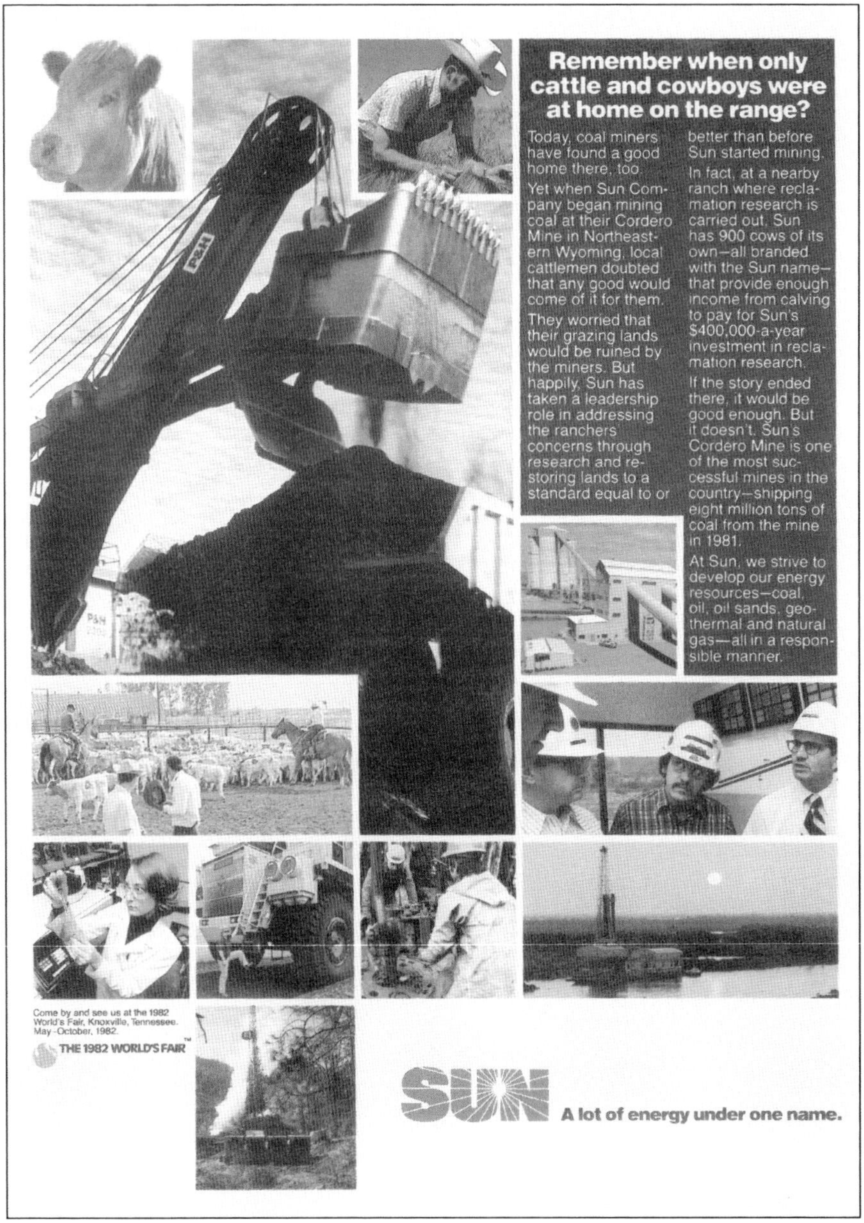

A full-page advertisement about Sun Company's Cordero Mine addresses early concerns about mining and ranching sharing the beautiful Wyoming prairie. In 1976, when Cordero Mining Company was looking for approval of its reclamation plan, Secretary of the Interior Thomas S. Kleppe made the decision to approve a 30-year plan of mining and reclamation for the 6,560 acres in the Sun Oil lease. The plan was for 337 million tons of coal to be taken from the federal lands, with only 3,800 acres of the leased land actually mined. Coal developments occurring in the Powder River Basin in 1976 were delayed due to a lawsuit known as *Sierra Club v. Morton*. The case had no impact on coal mines directly, other than slowing the leasing and permitting process. The issue was what steps the federal government needed to take in leasing federal coal reserves if that lease had a potential impact on the environment. To help those ranchers' peace of mind, an Environmental Impact Statement is now required as part of the mine permitting process.

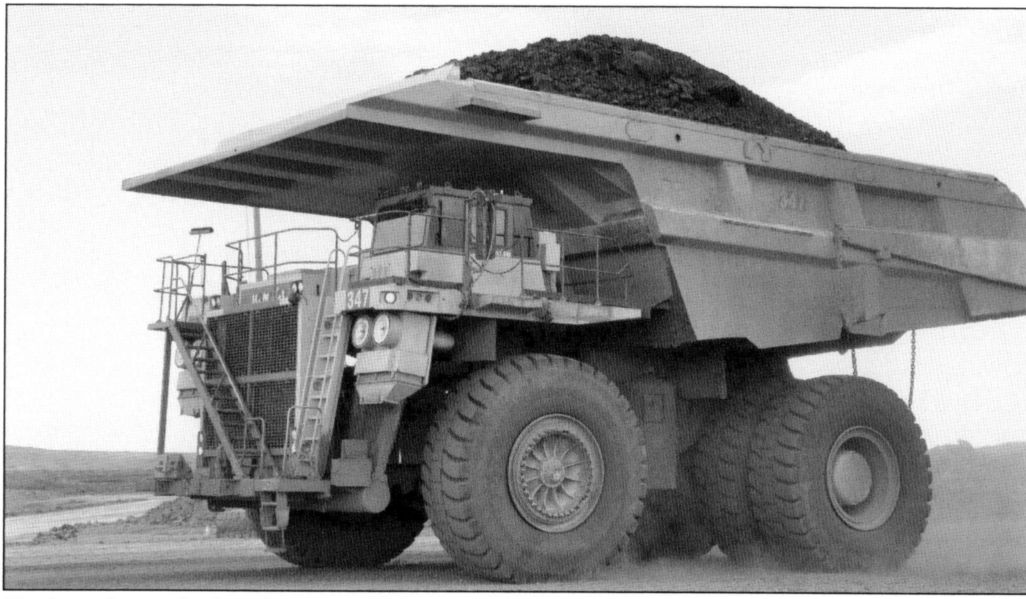

Haul trucks in the Powder River Basin started in the 1970s at approximately 170-ton capacity. Today there is a 400-ton class of haul trucks. The largest at the Cloud Peak mines are 340-tons made by Komatsu. (Courtesy of Cloud Peak Energy.)

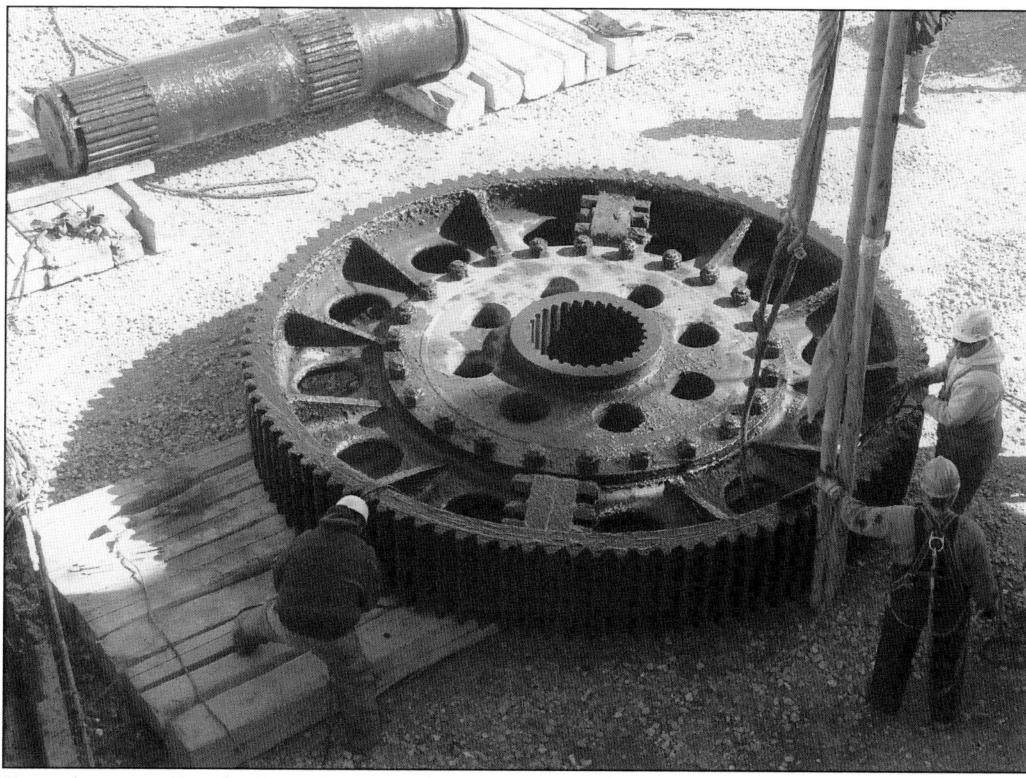

Everything at a Powder River Basin coal mine is huge. This photograph shows a swing gear from a shovel being repaired. An outage like this would require the shovel to be shut down temporarily. The swing gear operates or swings the bucket of the shovel. (Courtesy of Cloud Peak Energy.)

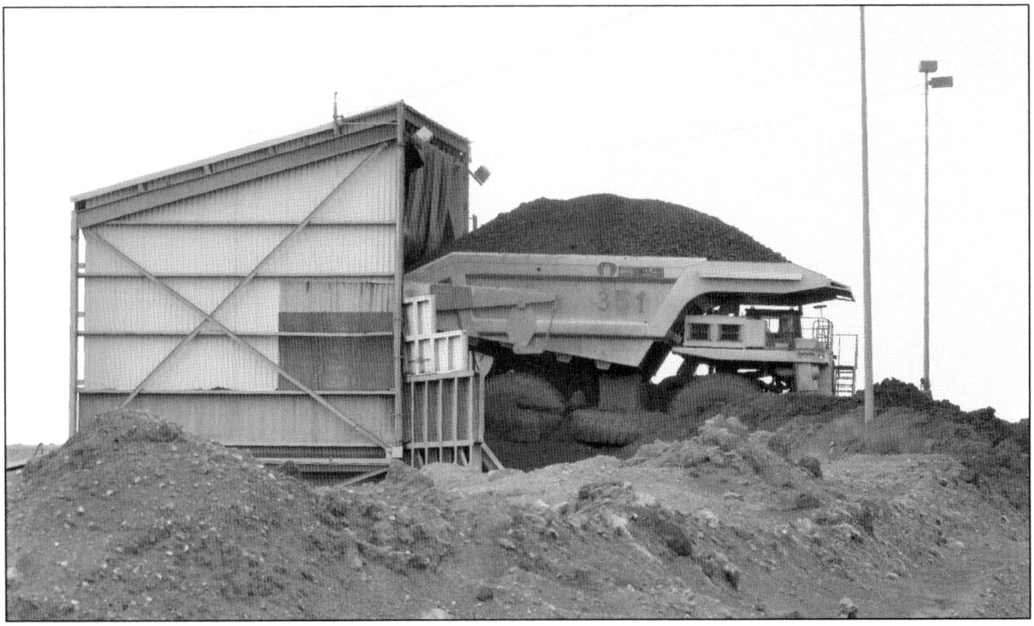

Haul trucks are loaded in the pit after a dragline strips overburden off the coal. At Antelope, the coal from the Anderson seam is 44 feet thick, and from the Canyon seam is 36 feet thick. At Cordero, the Wyodak seam ranges from approximately 55–70 feet in thickness. Fuel for machines like haul trucks is the highest commodity expense in mine operations. (Courtesy of Cloud Peak Energy.)

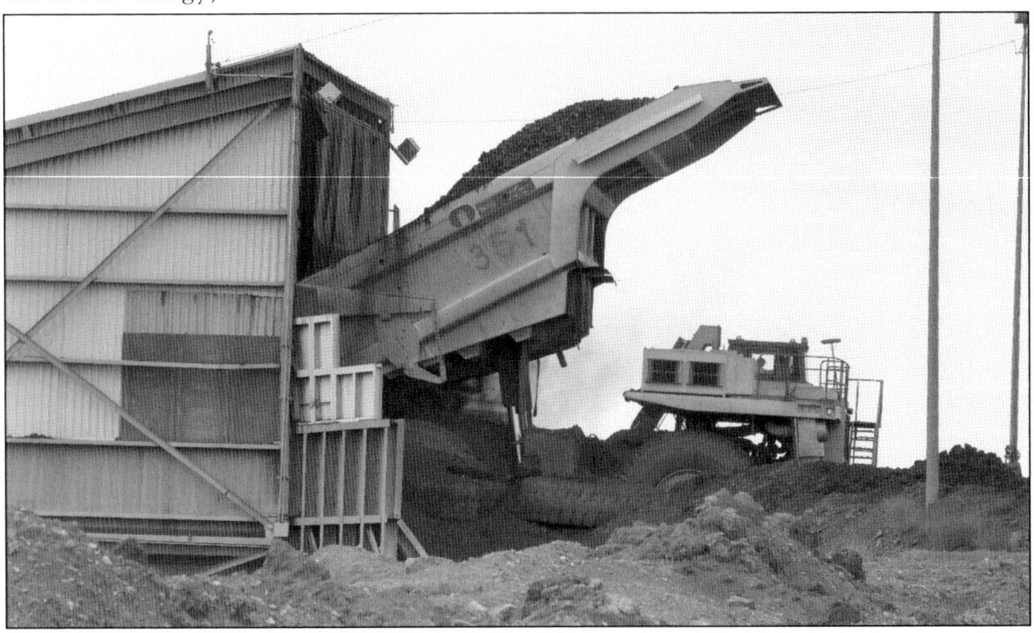

Loaded haul trucks dump their load into the hopper that is covered by a stilling shed to control any dust. Coal goes into the primary crusher, and then a secondary crusher. Customer requirements determine how small the coal is crushed, and whether it is blended with other coal to meet quality requirements. After the truck is emptied, it returns to the pit to be filled again and again. (Courtesy of Cloud Peak Energy.)

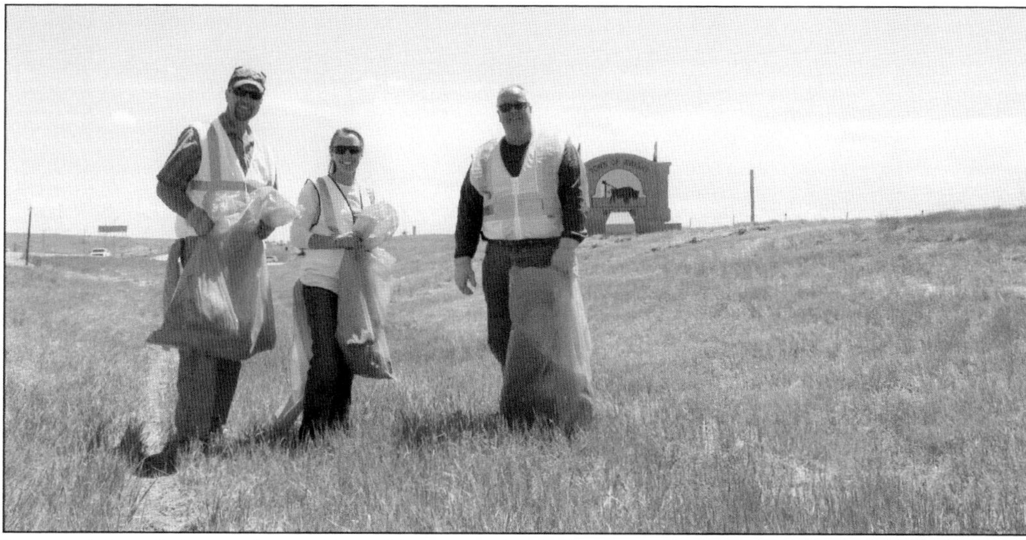

Pictured from left to right are Dave Traverso, Amy Clemetson, and Dave Dooley as they pick up trash in Wright, Wyoming. The communities in which Cloud Peak Energy has its mines and offices play a significant role in the company's success. Their employees live there, a large portion of their principal activities occur there, and they are neighbors to their operations. Their communities in the Powder River Basin of Wyoming include Douglas, Wright, Gillette, and Sheridan. (Courtesy of Cloud Peak Energy.)

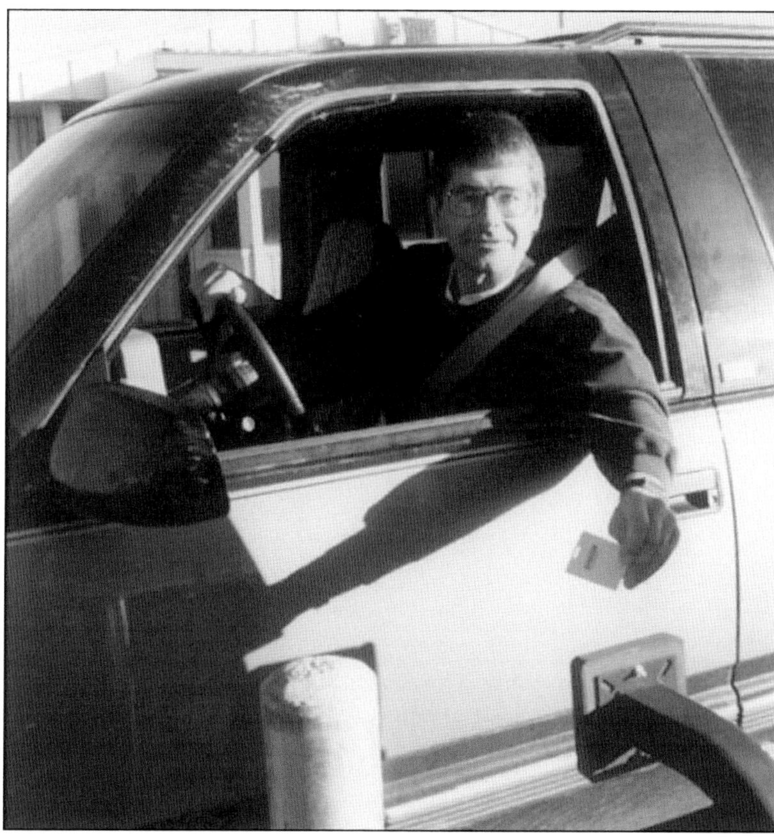

Dean Dvorak started his mining career as an engineer with the North American Coal Company in Bismarck, North Dakota. He joined Kennecott Energy's management team at Cordero Mine in 1994, advancing to general manager in 1995. Dvorak was promoted to CEO and president of Kennecott Energy in 2001. A much-loved friend and president was lost when Dean died in a car accident on June 29, 2002, near Reva, South Dakota. (Courtesy of Cloud Peak Energy.)

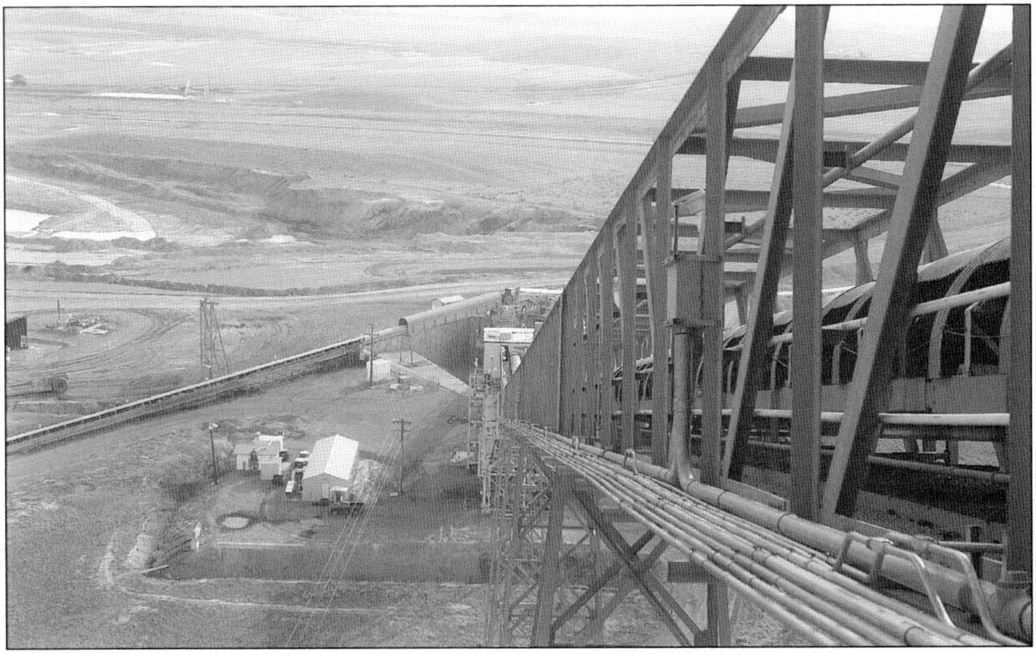

On average, coal travels roughly 2.5 miles from pit to crusher, depending on which pit at Antelope the coal comes from. This photograph shows part of the mine from the top of the conveyor. A very safe and productive mine with about 545 employees, Antelope celebrated six million man-hours without a single lost-time accident in May 2011. (Courtesy of Cloud Peak Energy.)

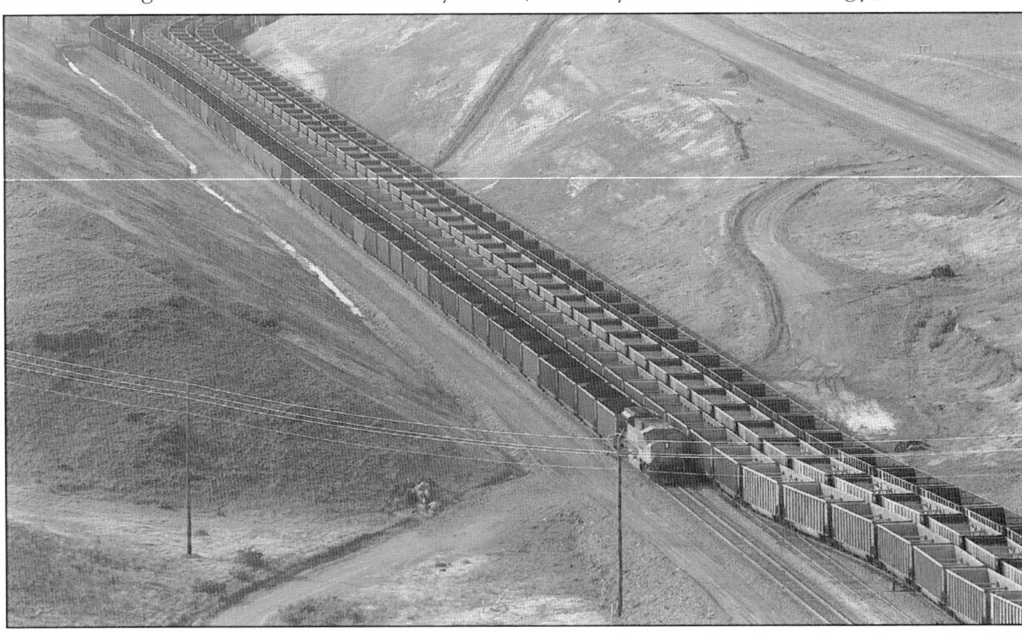

Antelope Mine has two dedicated inbound tracks, two dedicated outbound tracks, and a fifth track that can switch between inbound and outbound, but relatively little storage. According to mine manager Steve Cowan, "We have to have a steady stream of trains cycling through the loadout, otherwise we fill up in about four hours." The feed rate out of the silo allows them to load a train in approximately an hour and a half. (Courtesy of Cloud Peak Energy.)

Safety and tonnage milestones are celebrated at mines throughout the Powder River Basin. A banner is hung on the front of a train at Antelope Mine with the silo in the background. The milestone celebrated in this photograph is one billion tons of coal shipped safely to date from Kennecott Energy Company. (Courtesy of Cloud Peak Energy.)

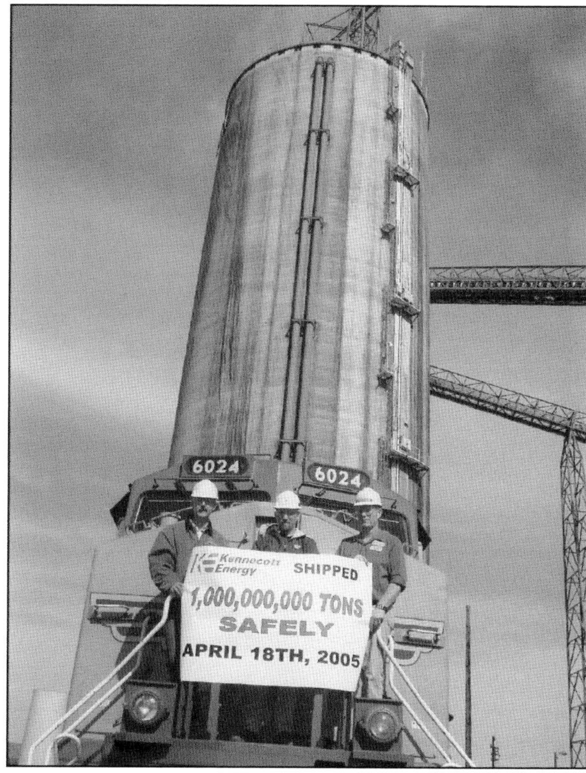

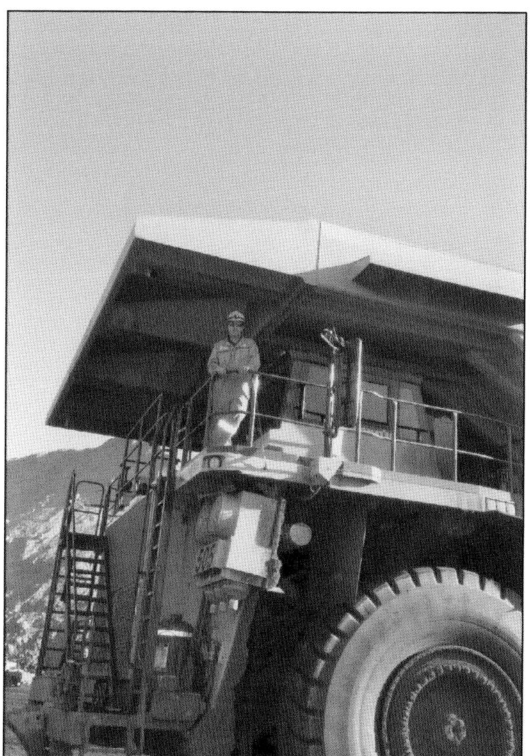

Without seeing it in person, the relative size of a haul truck tire is hard to imagine. Cordero Mine hosts a Family Day most years, when families of employees can visit the mine, ride in a haul truck, as well as enjoy games and refreshments. It's staggering for nonemployees to learn that a tire for this haul truck could cost in the range of $38,000 to $50,000. (Courtesy of Cloud Peak Energy.)

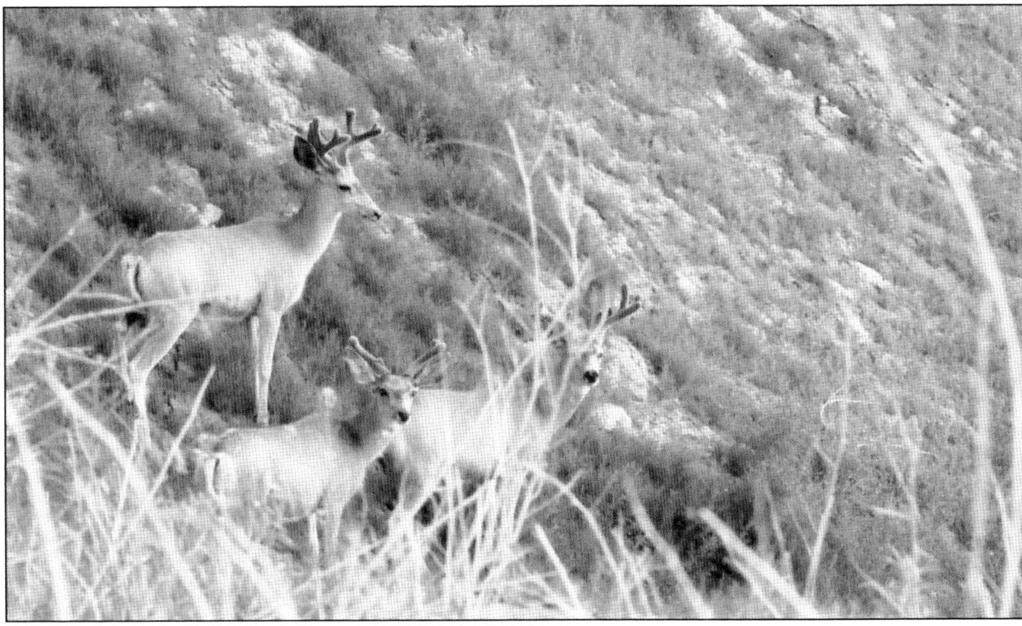

Mule deer live in a paradise of sagebrush, greasewood, and high grass on mine property. Hunting is not allowed, but how do the deer know that? The mule deer is named for its ears, which are large like those of a mule. In winter, mule deer forage on conifers, twigs, and shrubs, especially sagebrush. The deer's metabolism slows in winter and they become less active to survive with less food. (Courtesy of Cloud Peak Energy.)

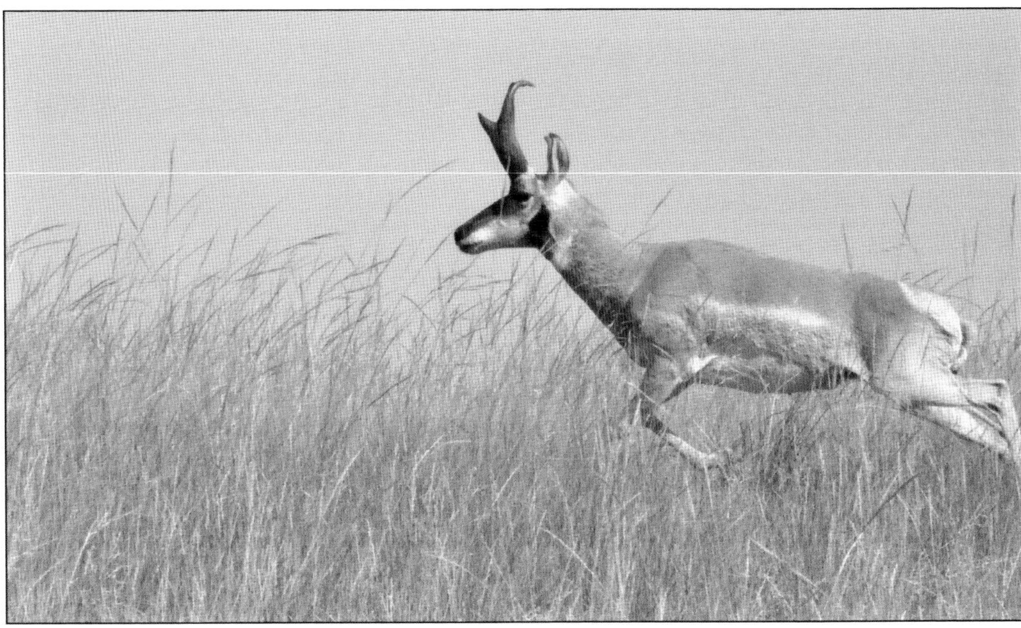

The pronghorn is not really an antelope but is often known in Wyoming as a pronghorn antelope, or simply antelope, since it closely resembles the true antelopes of Africa. It is often said there are more pronghorn than people in Wyoming. Mine property is a natural habitat for pronghorns, as their favorite diet consists of shrubs, grasses, and wildflowers such as prairie clovers, sweet clovers, and coneflowers. (Courtesy of Cloud Peak Energy.)

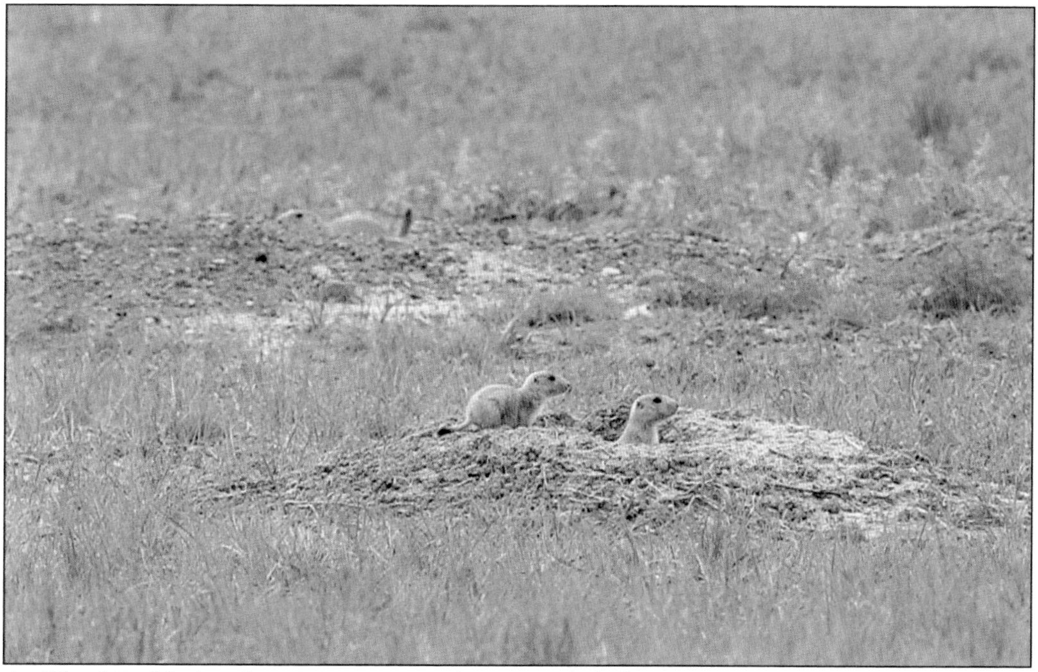

At Cloud Peak's Antelope Mine, prairie dogs provide habitat for the mountain plover, a small bird that likes to nest in disturbed areas like prairie dog colonies. The name mountain plover is a misnomer, as these birds like low, flat, dry land. (Courtesy of Cloud Peak Energy.)

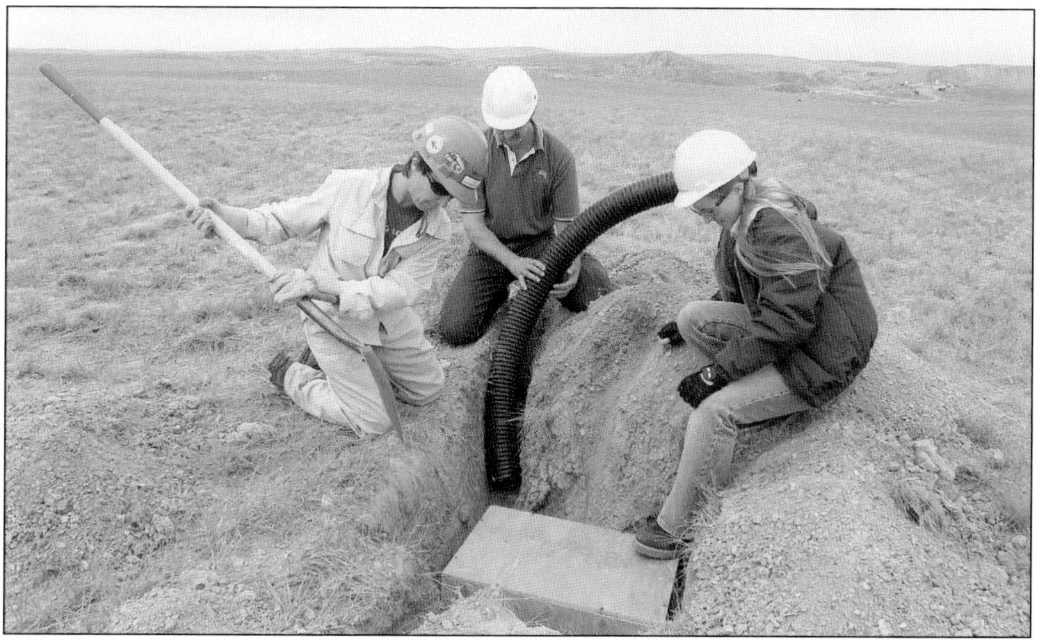

Antelope Mine employees place burrow boxes and entrances to relocate black-tailed prairie dogs onto reclaimed land. This helps establish mountain plover habitat and moves the prairie dogs out of the way of future mining. The boxes are necessary to keep badgers from digging out the prairie dogs. Cloud Peak works hard to ensure the cycle of life continues even during the mining process. (Courtesy of Cloud Peak Energy.)

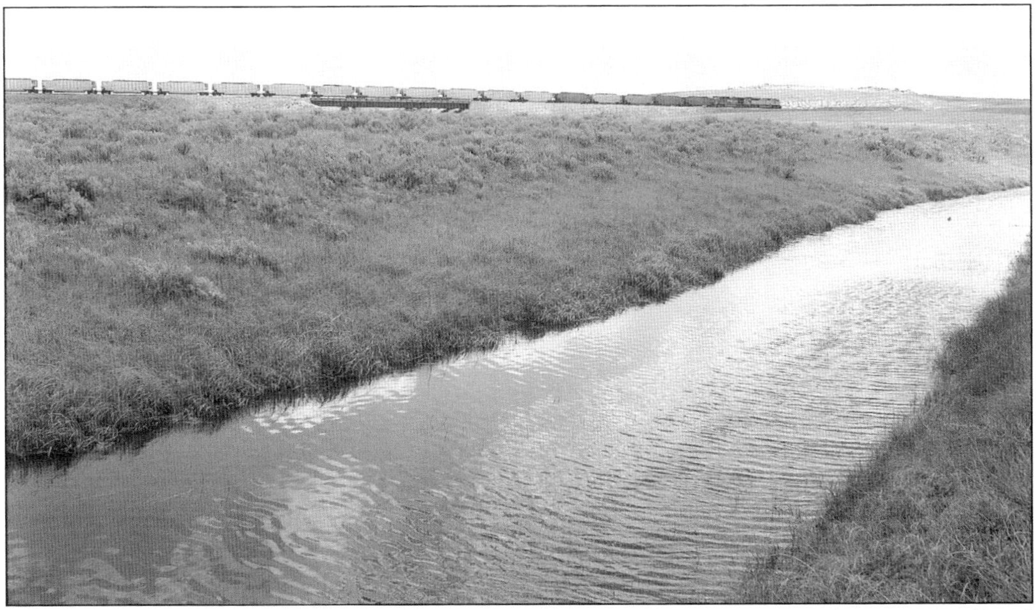

Mine environmental sites are audited and inspected regularly by Cloud Peak Energy staff and representatives of numerous agencies. This photograph shows an oxbow pond from the Belle Fourche River that was isolated by a very early diversion in the 1970s. When a river must be diverted, factors such as slope, land characteristics, chemical composition, the presence of fish, water quality, and vegetation are considered. (Courtesy of Cloud Peak Energy.)

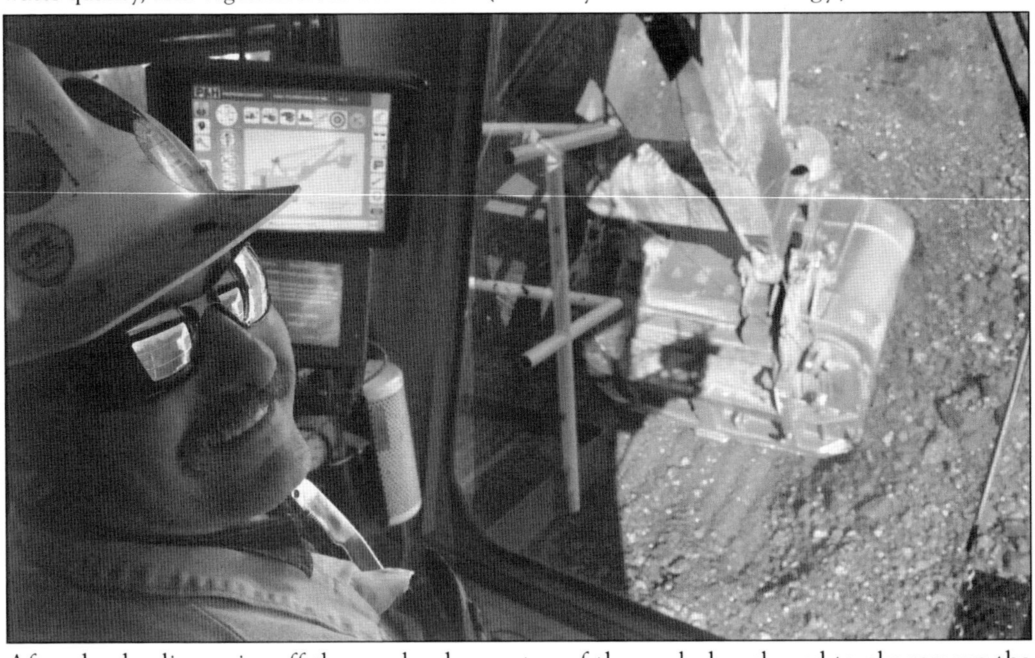

After the dragline strips off the overburden on top of the coal, shovels and trucks remove the coal for transport to the plant and loading at the silos. Cloud Peak Energy has six draglines and a large shovel fleet. One of the largest producers of coal in the United States, Cloud Peak Energy sells mainly to utilities and industrial customers, accounting for four percent of the electricity generated in the United States. (Courtesy of Cloud Peak Energy.)

Five

PEABODY ENERGY

Peabody Energy is the only coal company in the Powder River Basin to retain the same name since it began doing business in Campbell County. In fact, the Peabody empire dates back to 1883 in Chicago, when Francis Stuyvesant Peabody noticed residents were using more coal than wood to heat their homes. He believed coal was the rock that would build America, and he was right. He started Peabody, Daniels & Company with a partner, Edwin F. Daniels, and $100 in capital. He used a mule-drawn wagon to sell coal door-to-door. In 1885, he bought out his partner and became sole proprietor of the new Peabody Coal Company. During Peabody's early career, Thomas Edison revealed the coal-fueled electric generator, a machine that would forever change Peabody's future. The advent of the industrial age required power plants that generated energy for both businesses and homes. Peabody began purchasing reserves and opening mines to supply the need. His early purchases spanned seven states, including Wyoming. Fast forward to 1983, when Peabody celebrated its centennial. Relationships with customers were strong, productivity was high, and safety was a priority. Campbell County mines showed record profits. All was not rosy, however, as the country entered the mid-1980s. Computer technology, stiff competition, and interest in energy self-sufficiency were concerns during the Reagan administration. Anticipating huge reserves in the Powder River Basin, Peabody formed its Powder River Coal Company subsidiary well before Reagan's term in office. Increasing regulatory oversight in the 1980s soon increased demand for the West's low sulfur coal. The company began a course of action that would reap the rewards of years of prospecting. Peabody entered into a joint venture with oil company Panhandle Eastern Corporation to open the North Antelope Mine in the southern portion of the Basin in 1983. A year later, the company began developing the adjacent Rochelle Mine. In 1994, Peabody purchased the Caballo and Rawhide Mines in Wyoming from Exxon Coal USA, Inc. In 1999, the North Antelope and Rochelle Mines merged, creating Peabody's flagship, North Antelope Rochelle Mine.

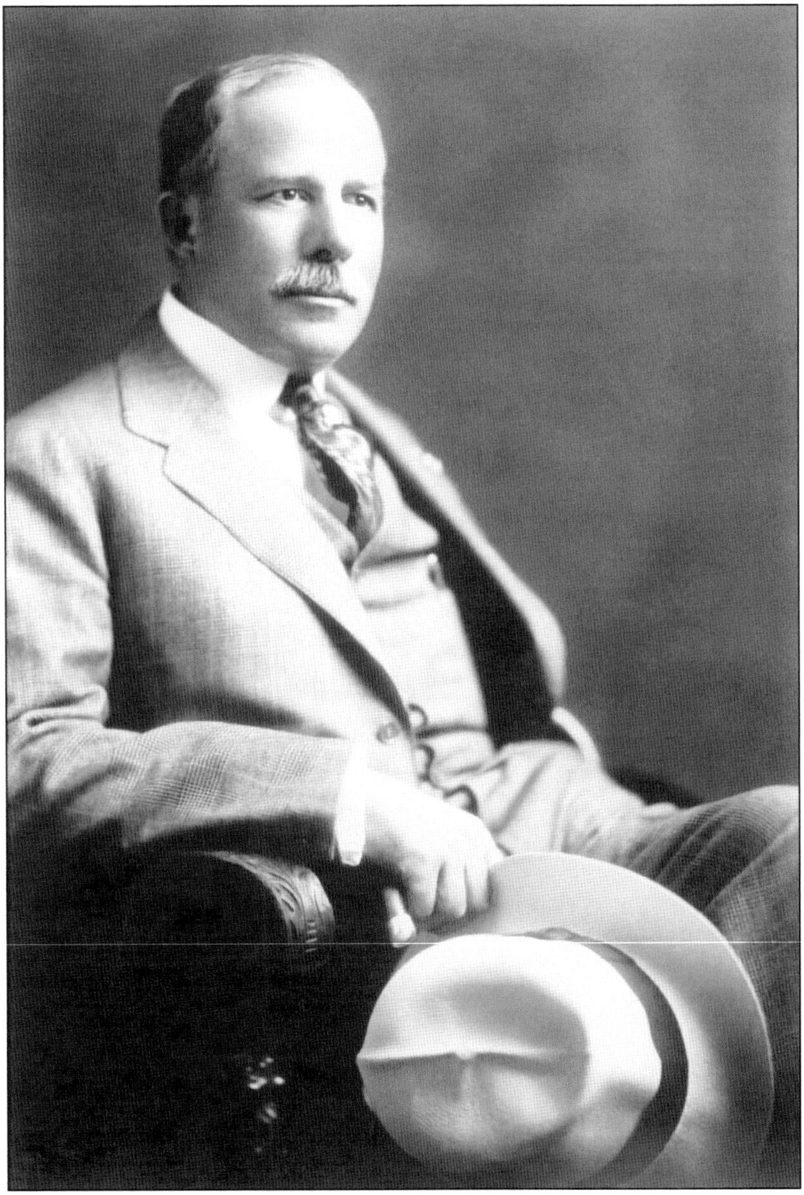

Francis Stuyvesant Peabody was born in 1858 in Chicago. His father, Francis Bolles Peabody, rose to prominence as a financier and attorney. His mother, Harriet Cutter Ten Broeck Peabody, was a genteel lady well prepared to make her way in Chicago society. When Francis was twelve, the Great Chicago Fire of 1871 burned 18,000 structures, including the Peabody family home. The family was resilient and returned to help rebuild Chicago. Following in his father's footsteps, Francis graduated from Yale University, but was not interested in being an attorney like his father. His timing in starting Peabody, Daniels & Company was excellent. Coal had surpassed wood as the primary energy source in America. New technologies and manufacturing methods were being discovered. Conveniences once reserved for only the wealthy, such as running water, steam heat, and indoor plumbing, soon became commonplace. Chicago's thriving economy was fueled by coal, and Francis was perfectly positioned to meet the need. By 1890, Francis was one of the wealthiest men in America at age 31. (Courtesy of Mayslake Peabody Estate and Peabody Energy.)

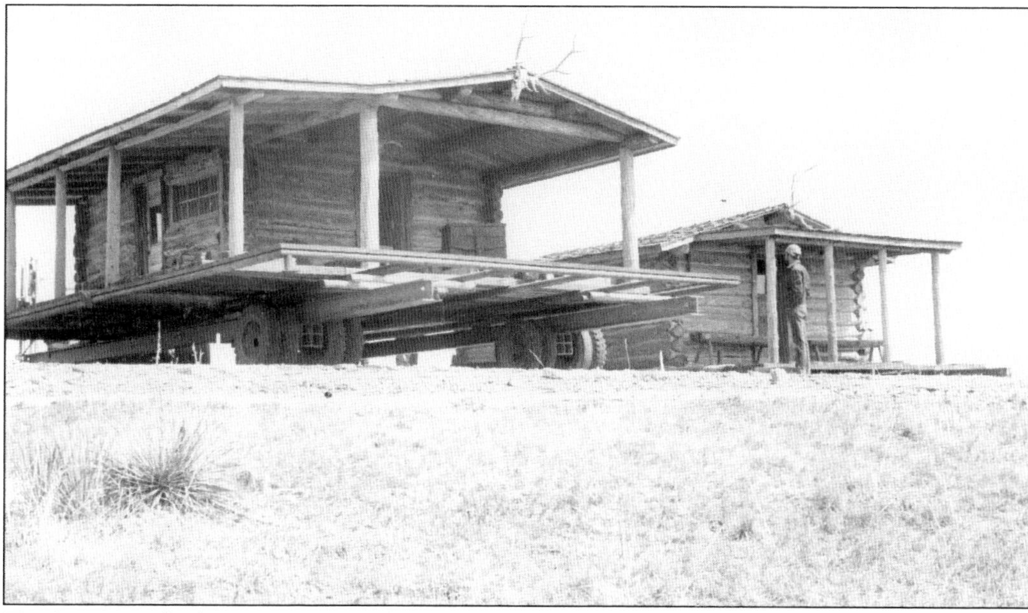

Two homestead cabins were restored and moved to Carter's Rawhide Mine to be used as a visitor's center in 1978. The older cabin was built in 1901 by L.A. Brown on his ranch near Weston. The cabin had a sod roof and cement chinking between the logs when Brown lived in the cabin with his wife, Vinnie. Restoration of the cabins was performed by Bob Edgar of Cody, Wyoming.

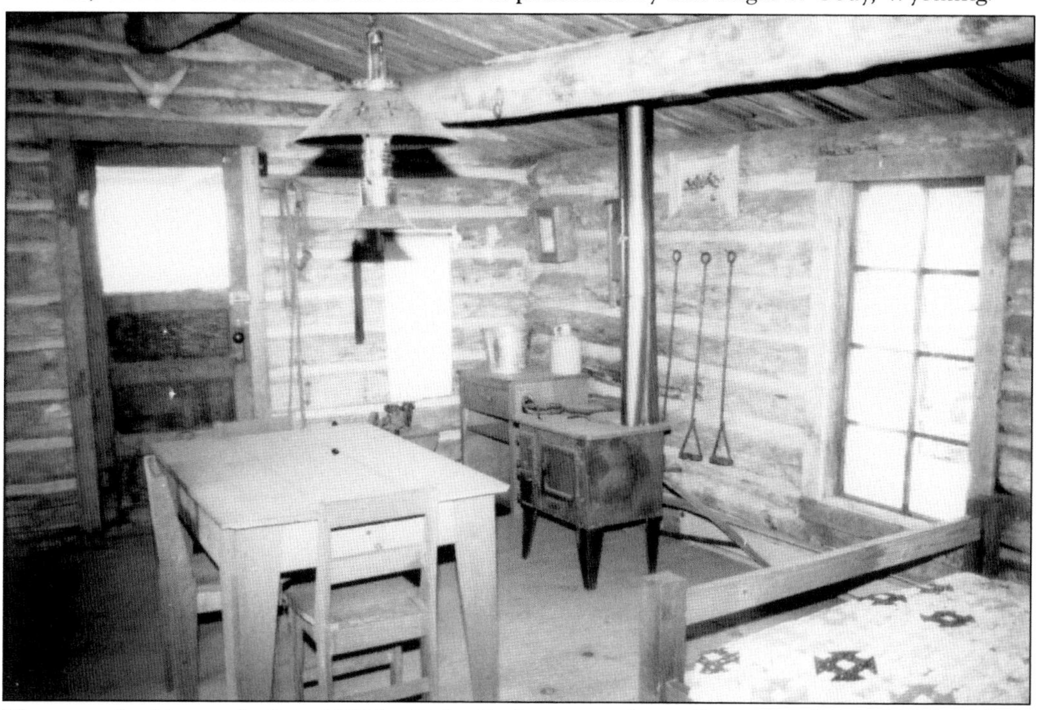

The second cabin was built in 1930 for Opha Suckow, a teacher at Bitter Creek School. The cabins were placed overlooking the pit at Rawhide Mine and later used as a museum and visitor's center. A newspaper advertisement showed that Rawhide Mine used to give guided tours by appointment. Rawhide Mine is approximately 10 miles north of Gillette.

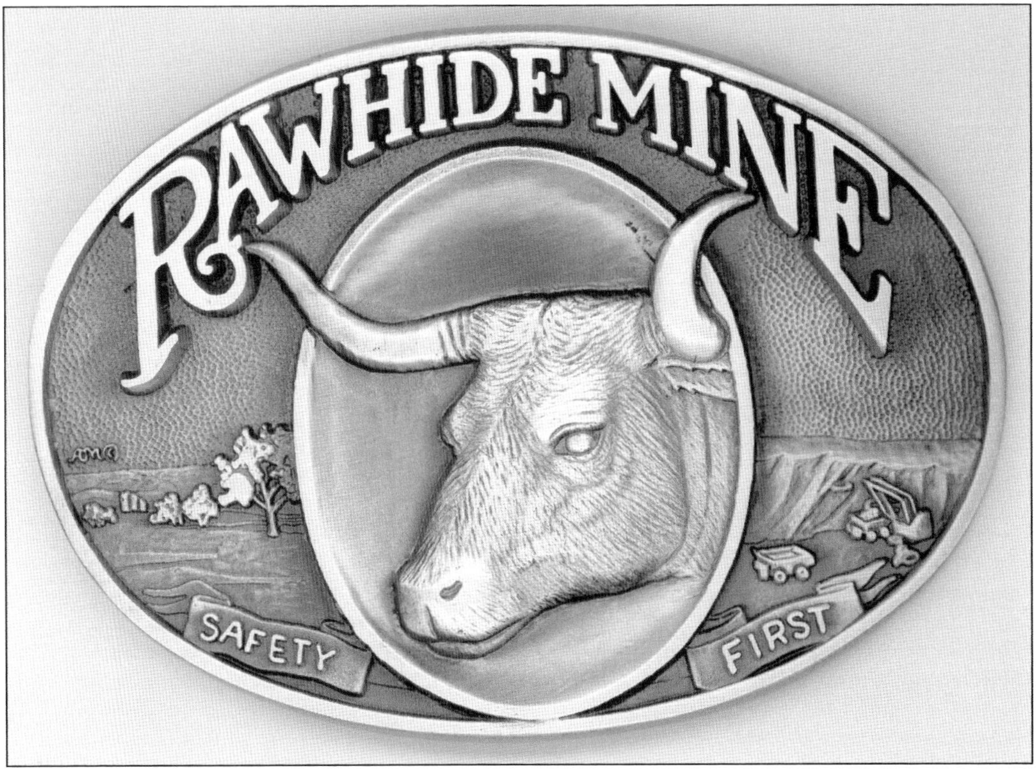

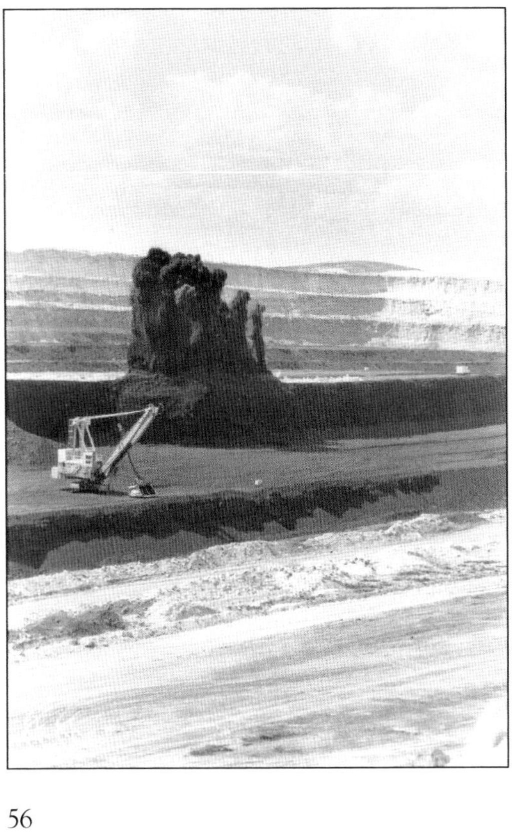

Rawhide Mine was dedicated in September 1977, with a crowd of 500 visitors on hand. C.E. "Ed" Smith Jr., president of The Carter Mining Company, assured the community they would be good neighbors and return the land to equal or better condition than at its origin. With a production goal of 12 million tons of coal a year, Rawhide would be producing electricity equal to 100,000 barrels a day of oil production.

Construction began on Rawhide Mine in September 1974. The mine opened in 1976, with production beginning August 1977. Production at the mine has been continuous since its opening, except for an 18-month period beginning in late 1999, when the mine was idled due to market conditions.

Dusti Poppelton poses as an elf in front of Peabody's entry in the 1996 Festival of Trees, which benefited the Campbell County Memorial Hospital. Donna Thorne, in the Human Resources Department, was instrumental in leading the decorating committee. Peabody continues, to this day, to be involved in the Festival of Trees in a big way. (Courtesy of Sandie Jones.)

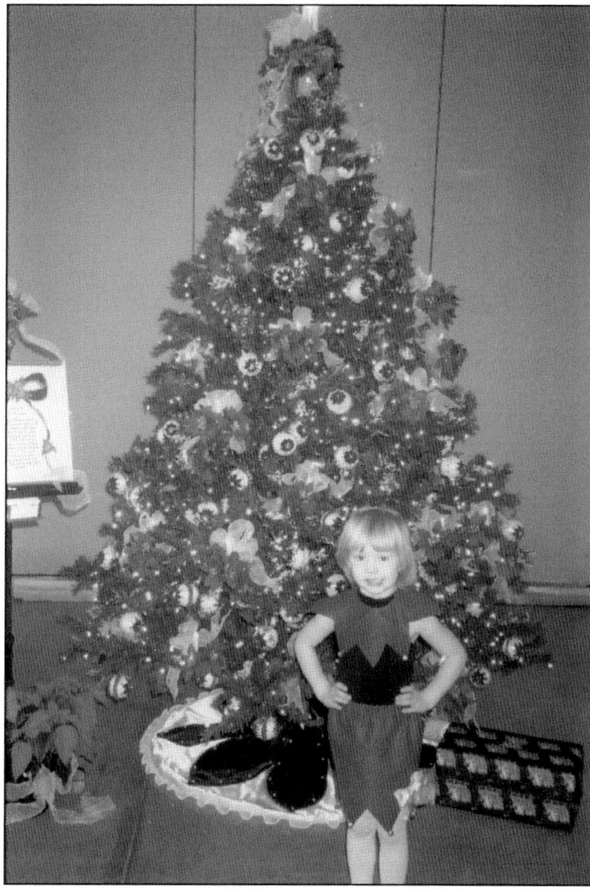

Carter Mining was committed to community involvement and employee fitness. This basketball team represented Carter Mining in the Campbell County Parks & Recreation league in the 1970s. Those shown are, from left to right, (first row) Bryan Leslie and Ken Fiebelkorn; (second row) Terry Rothwell, Nick Kasperik, Tom Kokel, and Terry Smotherman. Other sports sponsored by Carter included volleyball and trap shooting. (Courtesy of Sandie Jones.)

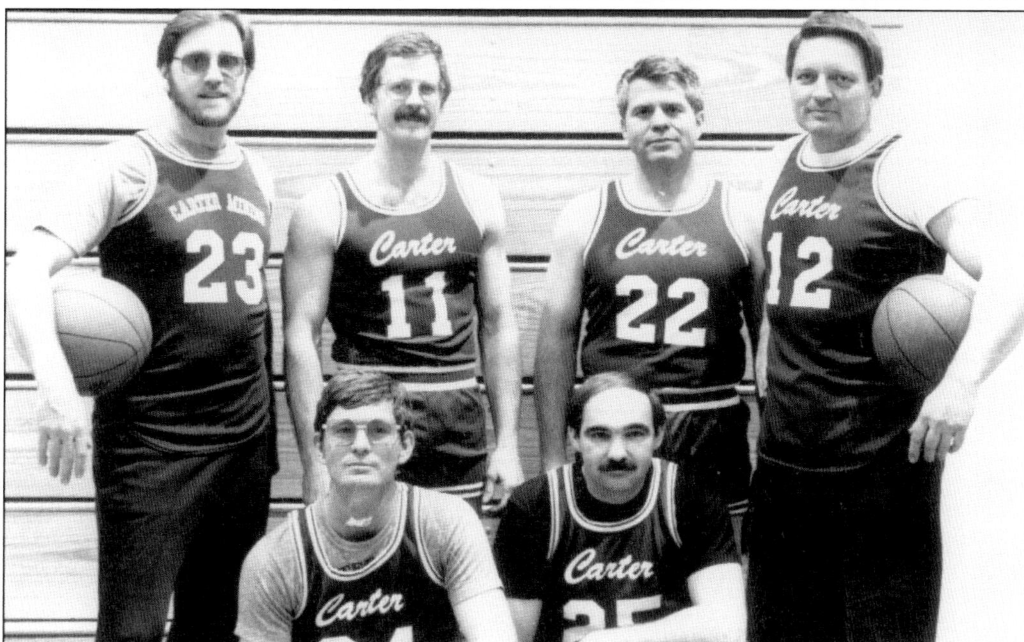

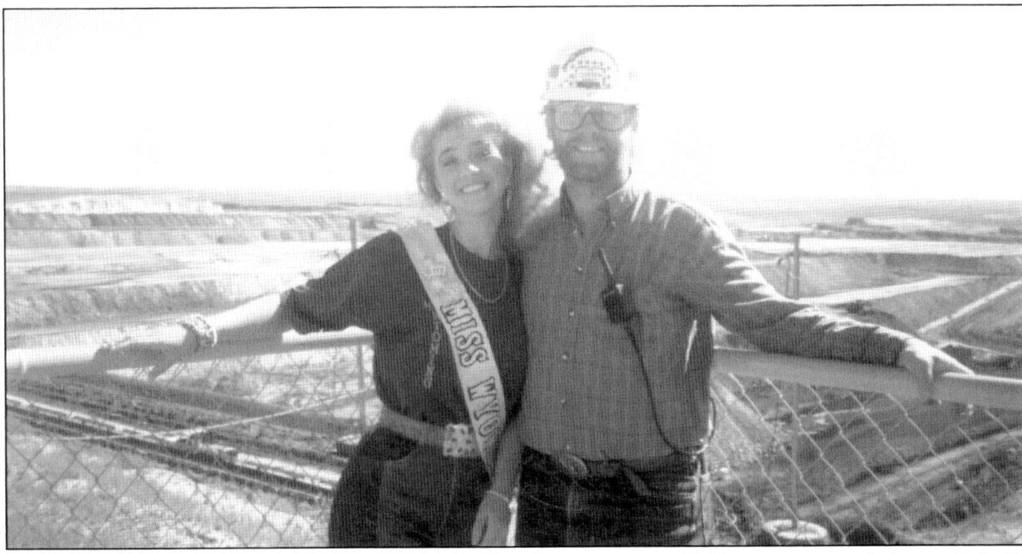

Lisa Postle, Miss Wyoming 1992, visited Caballo Mine with the 1993 pageant candidates. Caballo employee Craig Wilkinson readily volunteered as the tour guide. Photograph opportunities during the day included the popular pose by a haul truck tire and in front of the haul truck cab. (Courtesy of Sandie Jones.)

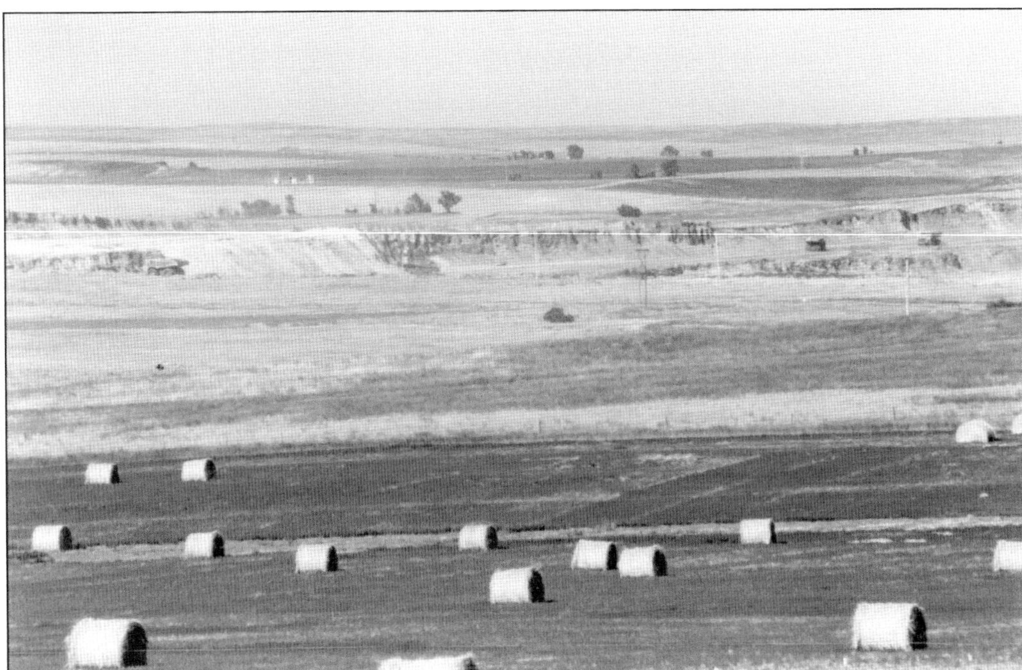

An independent contractor bales hay on reclaimed land at Caballo Mine with the pit in the background. The coal is mined from the Smith and Wyodak-Anderson seams, which average about 68 feet in thickness with an overburden of about 230 feet thick. The overburden is removed by shovel, with end-dump trucks used to haul the overburden to backfill areas that have been already mined.

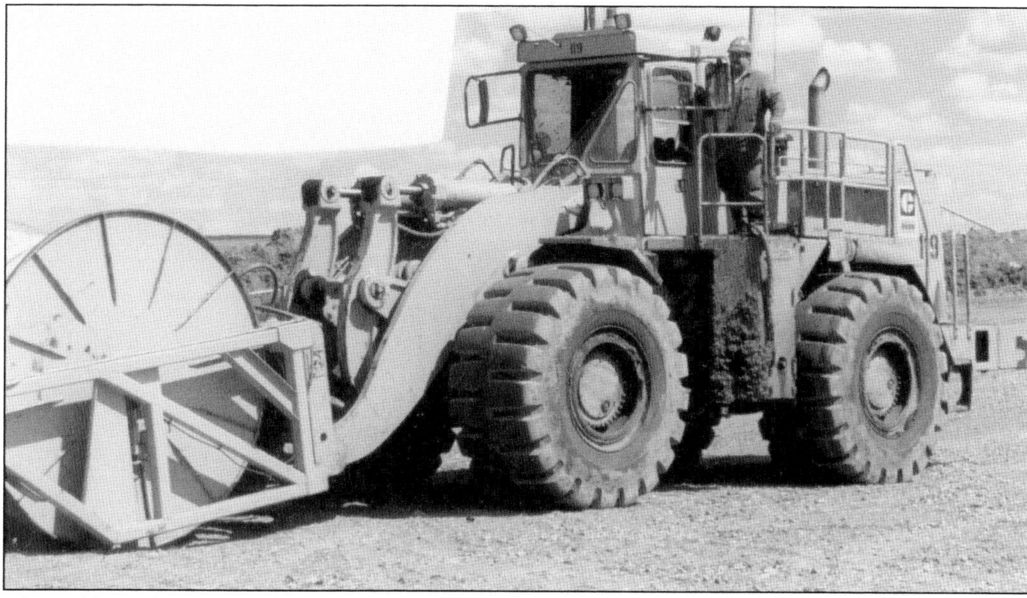

The cable reeler on the front of this dozer is an attachment that can be removed to put a bucket in its place. With the cable reeler on, it follows the huge shovel and rolls up electric cable when not needed. (Courtesy of Sandie Jones.)

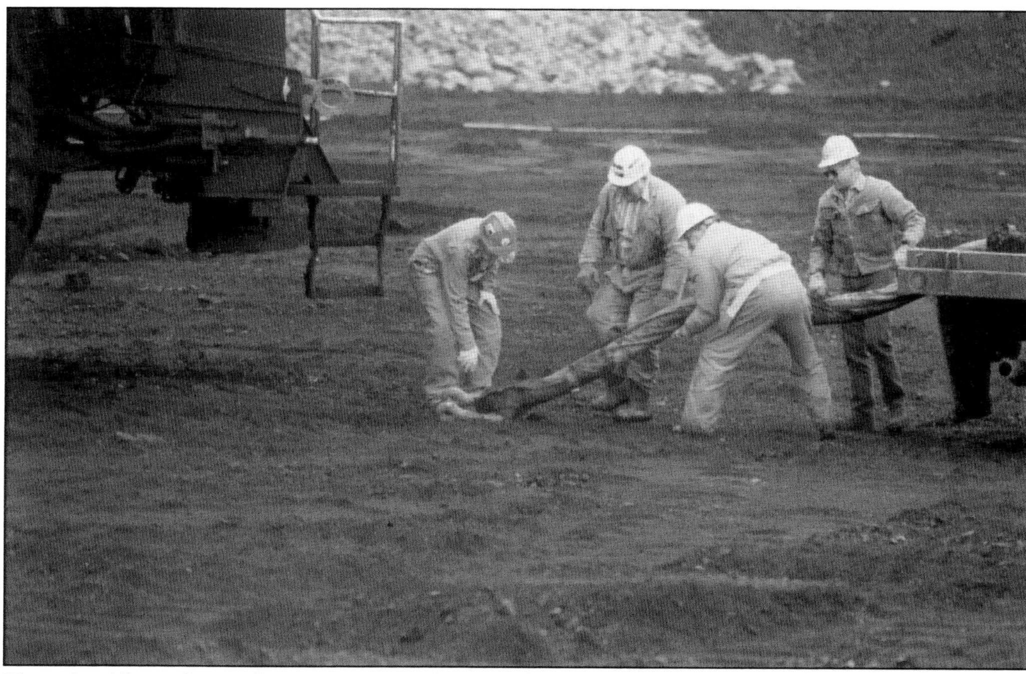

Shovel cable is shown here, requiring four employees to move it. They wear special gloves to insulate against the 4,160 watts in the cable. (Courtesy of Sandie Jones.)

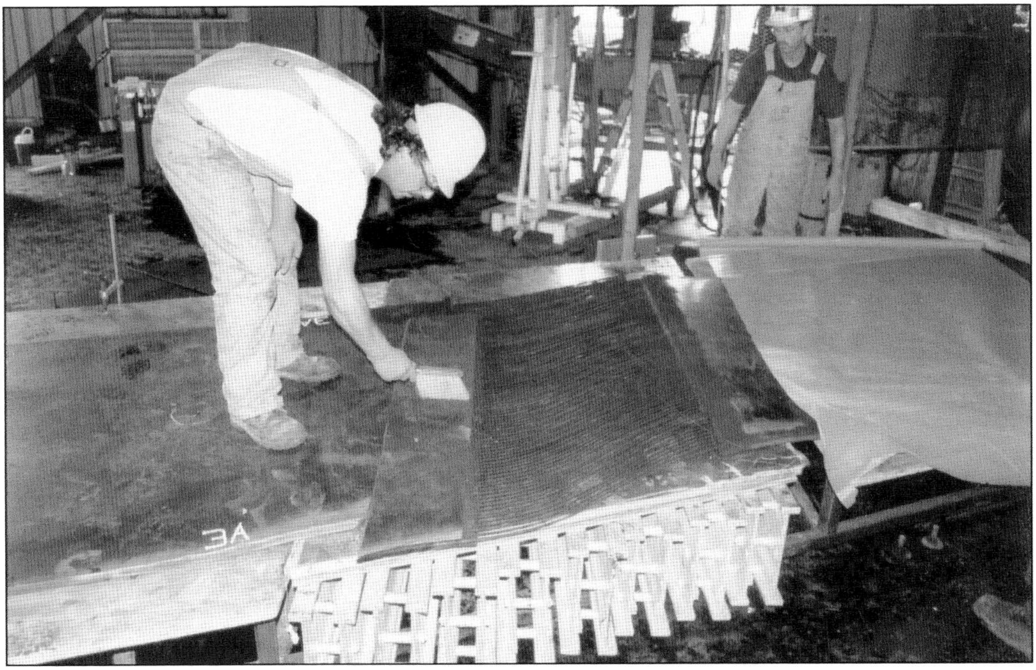

Inside the storage silos, a specialized vendor repairs the conveyor belt. The conveyor belt can be spliced many times when needed, rather than replaced. Caballo Mine has four storage silos fed by a 60-inch conveyor belt. (Courtesy of Sandie Jones.)

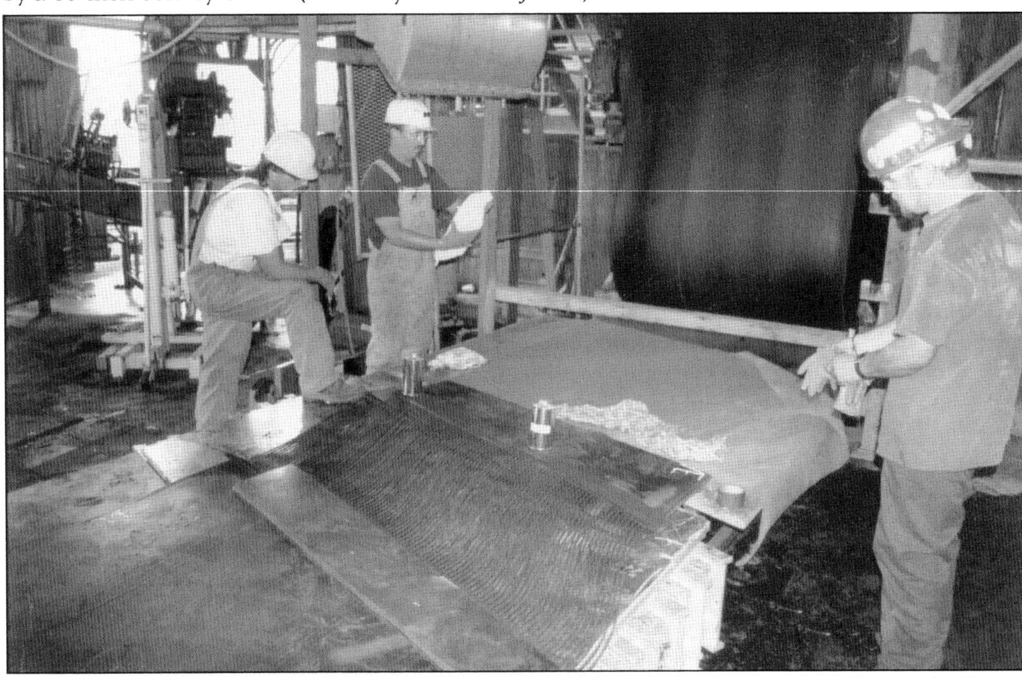

North Antelope Rochelle Mine is one of the largest coal mines in America, producing the lowest sulfur coal in the nation. The complex shipped a record 105.8 million tons in 2010 and more than 1.3 billion tons since the mine began. The mines operate on two 12-hour shifts per day, 365 days per year. (Courtesy of Sandie Jones.)

Carter Mining produced a newsletter called *The Carter Scoop* for employees, giving updates, production information, and employee spotlights. AMAX had a newsletter called *Mine Run*; Peabody has the *Pulse*. With over 500 employees between Caballo and Rawhide Mine, newsletters provided a unifying voice. (Courtesy of Sandie Jones.)

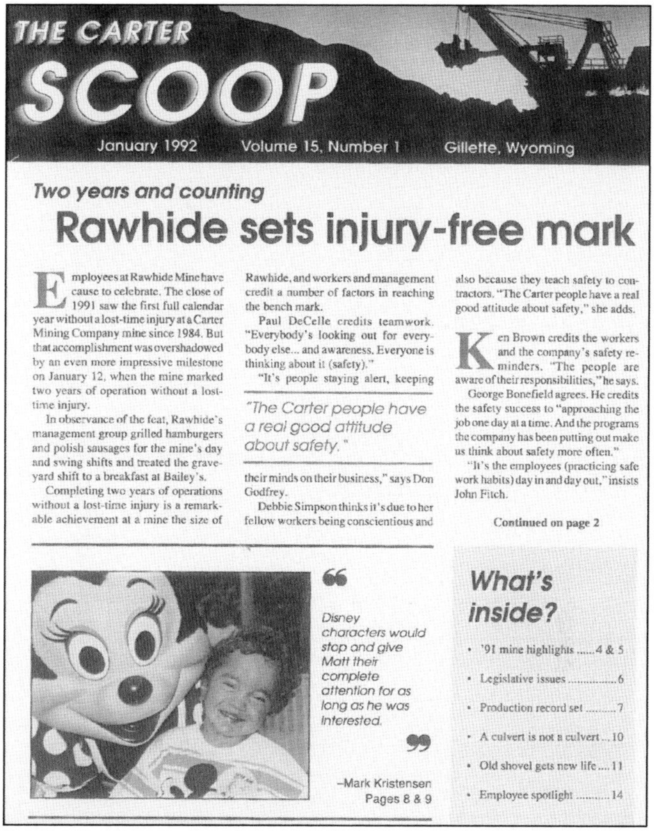

THE CARTER SCOOP

January 1992 Volume 15, Number 1 Gillette, Wyoming

Two years and counting

Rawhide sets injury-free mark

Employees at Rawhide Mine have cause to celebrate. The close of 1991 saw the first full calendar year without a lost-time injury at a Carter Mining Company mine since 1984. But that accomplishment was overshadowed by an even more impressive milestone on January 12, when the mine marked two years of operation without a lost-time injury.

In observance of the feat, Rawhide's management group grilled hamburgers and polish sausages for the mine's day and swing shifts and treated the graveyard shift to a breakfast at Bailey's.

Completing two years of operations without a lost-time injury is a remarkable achievement at a mine the size of Rawhide, and workers and management credit a number of factors in reaching the bench mark.

Paul DeCelle credits teamwork. "Everybody's looking out for everybody else... and awareness. Everyone is thinking about it (safety)."

"It's people staying alert, keeping their minds on their business," says Don Godfrey.

"The Carter people have a real good attitude about safety."

Debbie Simpson thinks it's due to her fellow workers being conscientious and also because they teach safety to contractors. "The Carter people have a real good attitude about safety," she adds.

Ken Brown credits the workers and the company's safety reminders. "The people are aware of their responsibilities," he says.

George Bonefield agrees. He credits the safety success to "approaching the job one day at a time. And the programs the company has been putting out make us think about safety more often."

"It's the employees (practicing safe work habits) day in and day out," insists John Fitch.

Continued on page 2

"Disney characters would stop and give Matt their complete attention for as long as he was interested."

–Mark Kristensen
Pages 8 & 9

What's inside?

In 1994, Caballo Coal Company, a subsidiary of Peabody Energy Company, purchased Exxon Coal USA, Inc.'s Carter Mining for $356.2 million. The purchase included Rawhide and Caballo Mines. This office is located at 505 South Gillette Avenue, today the headquarters of Cloud Peak Energy.

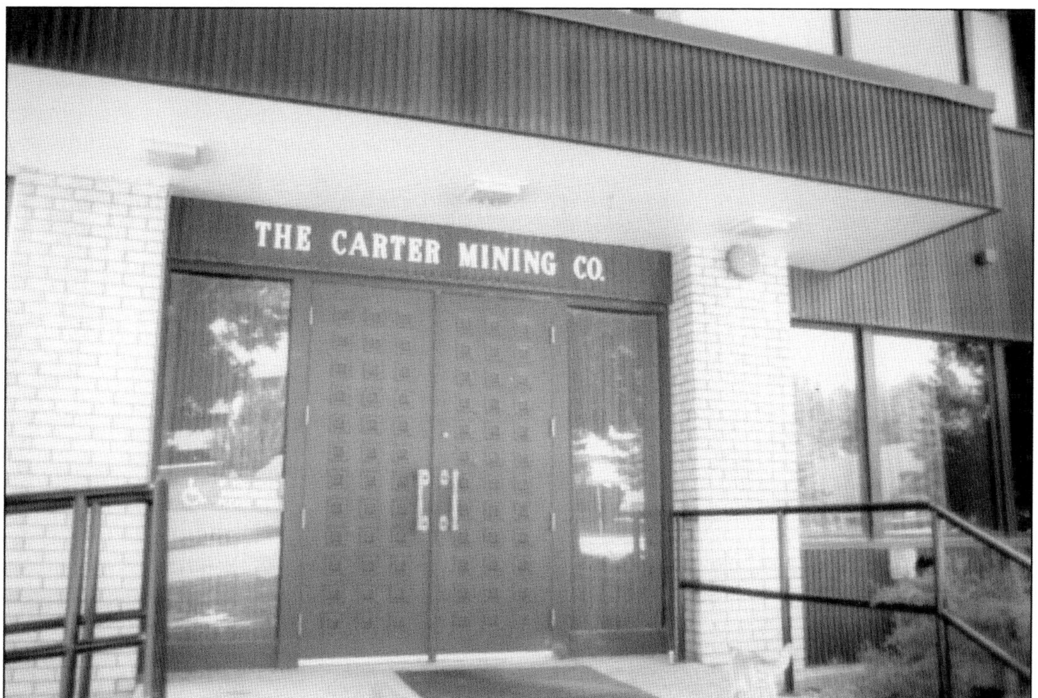

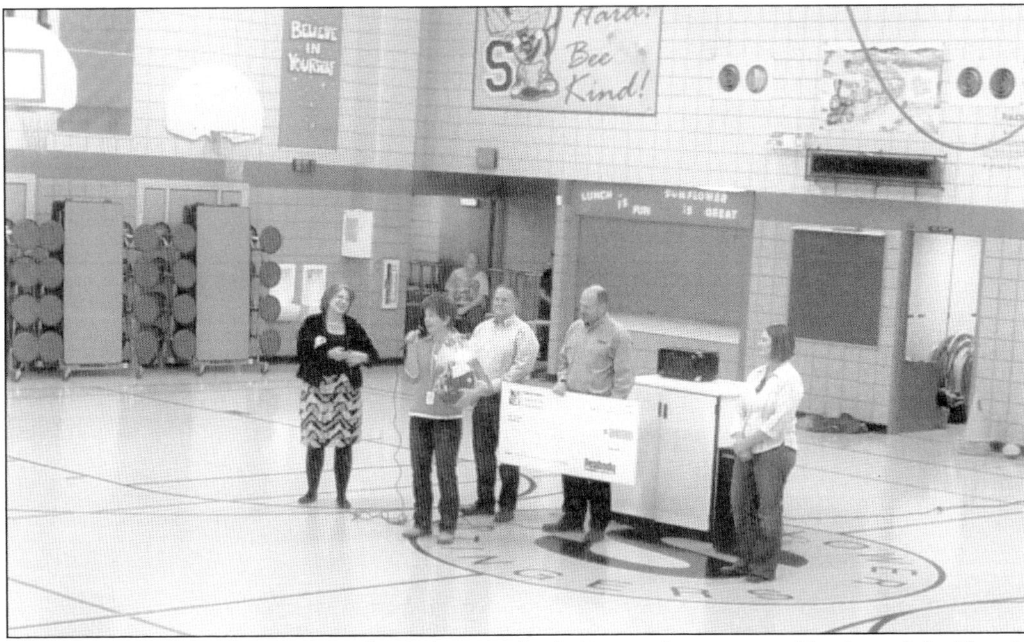

Kathy Kintz receives an oversize check from Peabody Energy for winning Teacher of the Year in 2012. Kathy also won the Arch Coal Teacher Achievement Award. With 25 years of experience, Kathy teaches third grade at Sunflower School in Gillette. Nominations for both awards are made by the public, and selection is made by a panel of the teacher's peers. (Courtesy of Kathryn Kintz.)

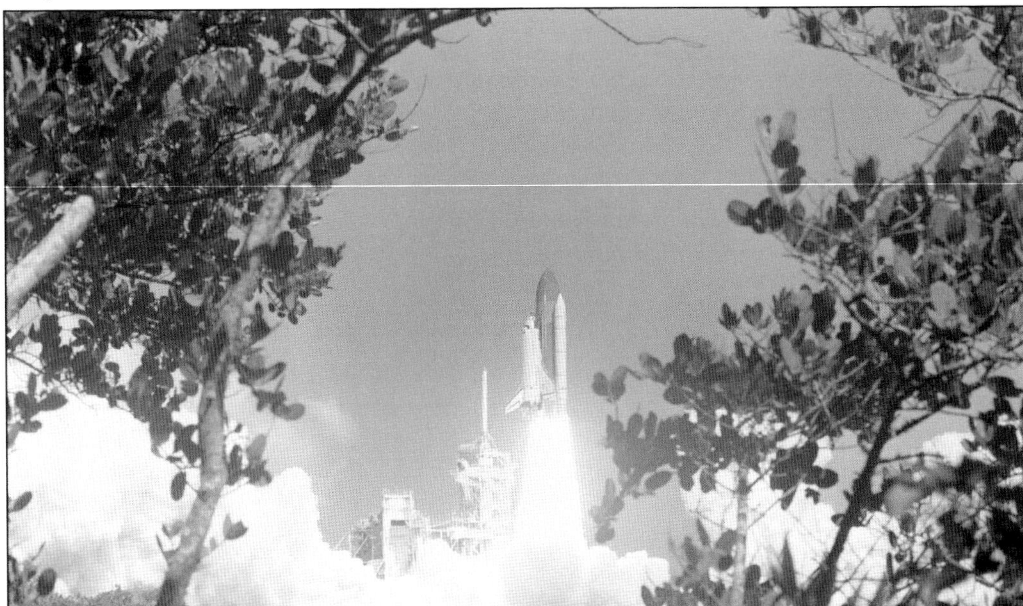

Peabody's Powder River Basin coal is truly "out of this world." Coal from North Antelope Rochelle Mine rocketed to outer space on the US Space Shuttle Discovery in 2001. The shuttle carried materials for experiments designed by 300 elementary-age students from St. Louis. The students hoped that the extreme temperatures of space would turn the coal into diamonds; this, unfortunately, did not happen.

Six

Wyodak

Of all the coal mines and coal companies in the Powder River Basin, Wyodak has the richest local history. It is the longest continuously operating surface mine in the United States. One part of the story began as a small underground mine in 1922, and another began as a surface mine with several unique features. The Homestake Mining Company, headquartered in Lead, South Dakota, purchased lands adjoining the Peerless Mine and created the Wyodak Mine to supply power to the Homestake Gold Mine in Lead. A small power plant was built shortly after the mine opened. The Wyodak Mine is a surface, or open-pit, mine close to the location of the Peerless. At some point in its history, Wyodak mined through the Peerless, but not before salvaging old rail track and other artifacts. The power plant near the mine was fed coal directly from the mine by a conveyor belt. This type of operation is called a "mine-mouth" operation. In 1923, the little village of Wyodak was created on mine property. When bad weather hit, the residents were available to mine coal to supply the power plant, and keep electricity going to Gillette. Homestake operated the plant from 1923 to 1956, when it sold the mine and plant to Black Hills Power & Light. The original brick building of the Homestake plant now houses the Neil Simpson Station and is still in use by the company, although it is scheduled to be shut down in 2013. The plant was the first to use an air-cooled condenser rather than water-cooled like most power plants. The mine is now operated by Wyodak Resources Development, a subsidiary of the Black Hills Corporation.

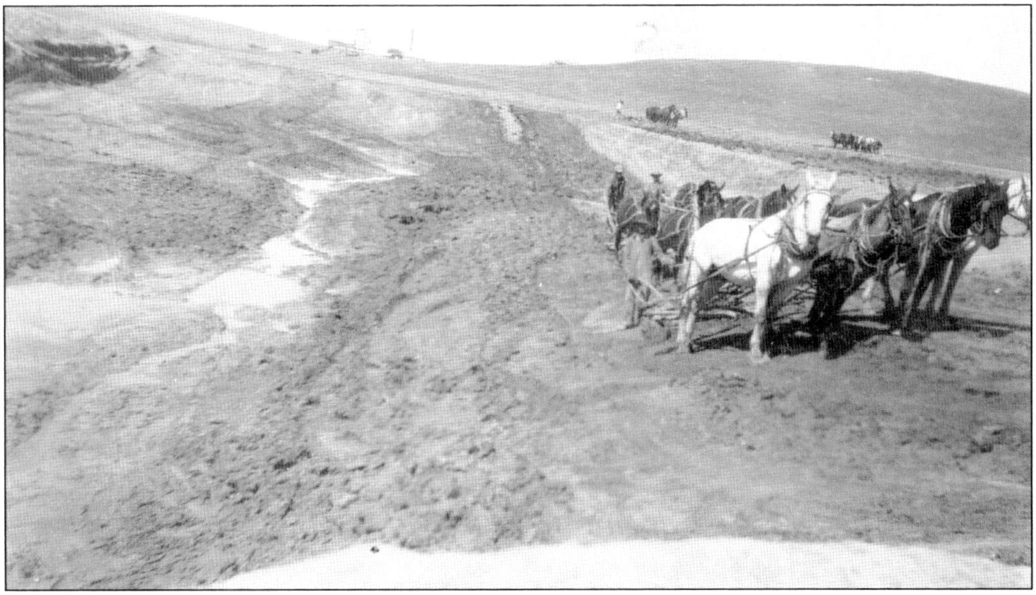

Homestake purchased the Wyodak property in 1921. From 1922 to 1924, horses and half-yard scrapers, called "fresnoes," were used to uncover the first small patch of coal. However, that method of stripping was too slow, especially after water and mud were reached. Many fresnoes used four-horse teams for maximum power.

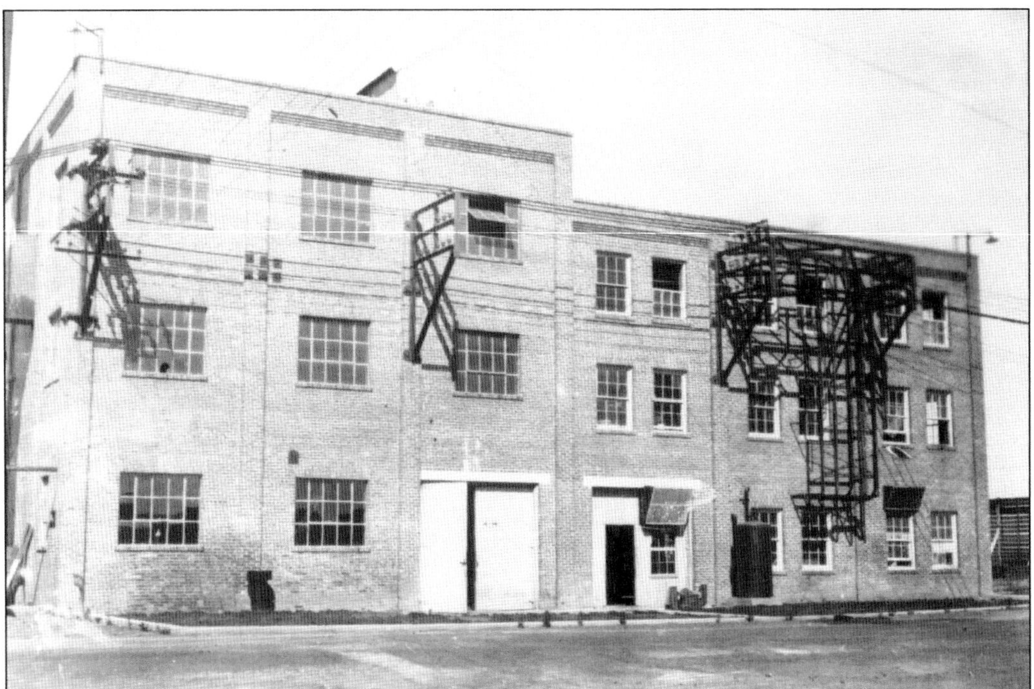

This photograph shows the main generating plant at Wyodak in the 1930s or 1940s. At that time, 120,000 tons of coal were taken annually from the pit. Sixty percent went to furnish fuel for the Homestake Mine in Lead, while the rest was used by the City of Gillette to create electricity or by the Chicago, Burlington & Quincy Railroad.

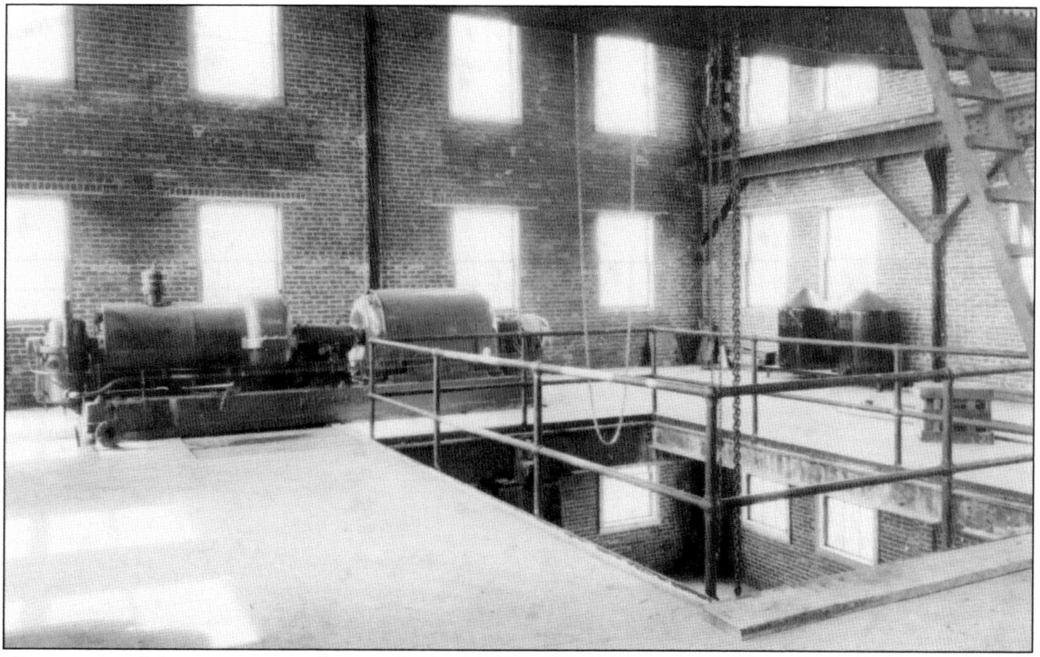

The inside of the generating plant is shown here on June 25, 1928. What appears to be a generator is seen at the left, with an opening to the next floor below shown to the right. There is also a ladder to the next floor above on the right. Natural light from multiple windows shows how clean the interior of the plant is, a standard that continues today.

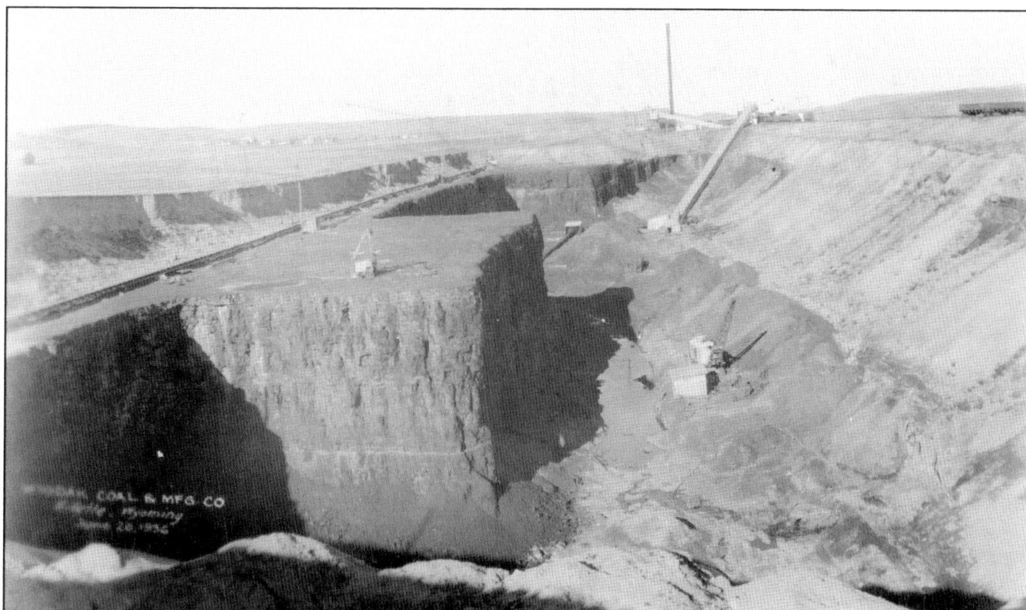

The size of the pit and extent of the overburden on top of the coal is displayed in this photograph. A block of coal remains on the left, soon to be blasted. Holes have already been drilled to hold the explosives. Note the conveyors in the pit that lead to the main conveyor, lifting the coal up to the plant. Wyodak was the first coal mine in the Powder River Basin to use conveyors to move coal to the plant, eliminating the need for coal haul trucks.

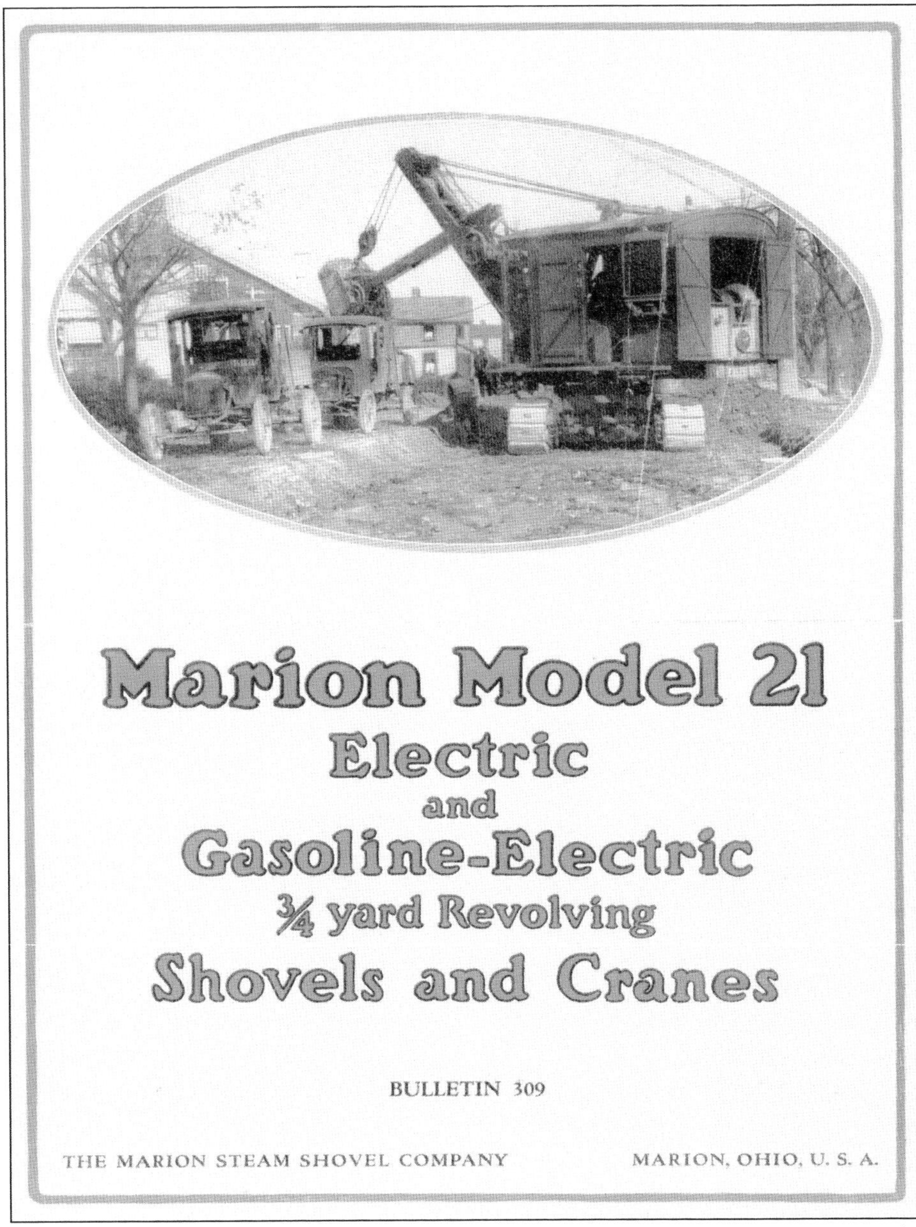

Many Marion shovel models were offered as steam-powered, then steam or electric, and finally only electric. In a 1929 presentation to the Hazelton (PA) Mining Institute, H.T. Gracely, Marion's general sales manager, said: "It is remarkable that a coal mining tool should not be a coal using machine. But the economies of electric shovels in this service are so great that the sale of a steamer is now quite unusual." The impact of coal mining has increased the assessor's valuation of Campbell County since the days when Wyodak was the only large commercial mine. The 1939 valuation was $8,475,718; in 1979, after several other large mines had started production, the valuation was $596 million. The 1987 valuation was $1,051,976,173; and with a substantial jump, the valuation in 2012 was $5,839,065,491. Not only did coal benefit the county tax coffers, but increases were significant over the years for oil, gas, railroad, and, at times, livestock. (Courtesy of Marion County Museum, Marion, Ohio.)

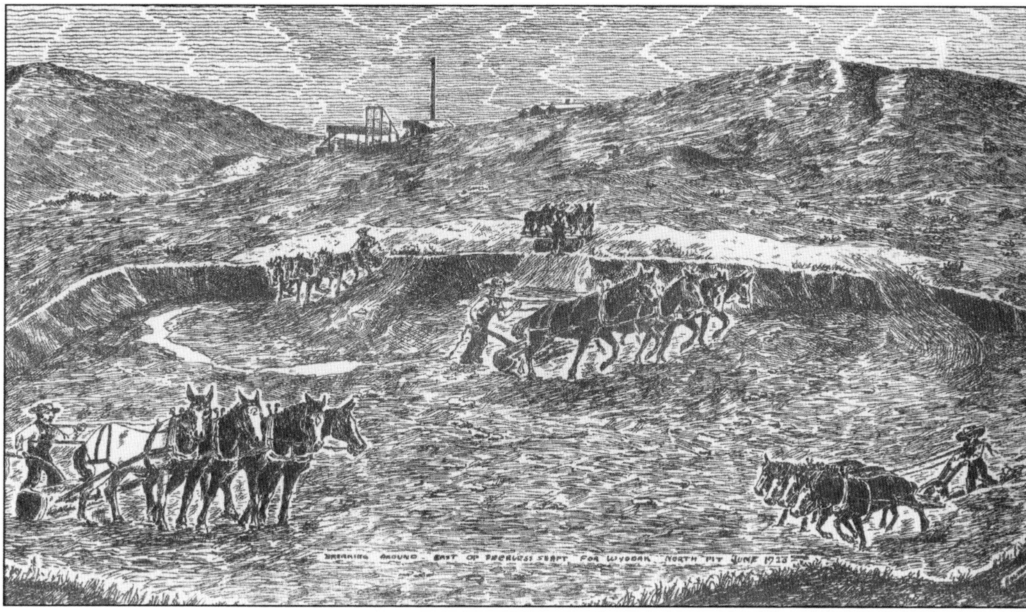

Local artist William Halvor Johnson made several drawings of Peerless and Wyodak Mines using pen and ink. This drawing from July 1, 1985, is titled "Breaking Ground East of Peerless Shaft for Wyodak North Pit June 1928." Johnson started working at Wyodak April 4, 1937, as a car loader under the tipple. Sometime later, he worked in blasting.

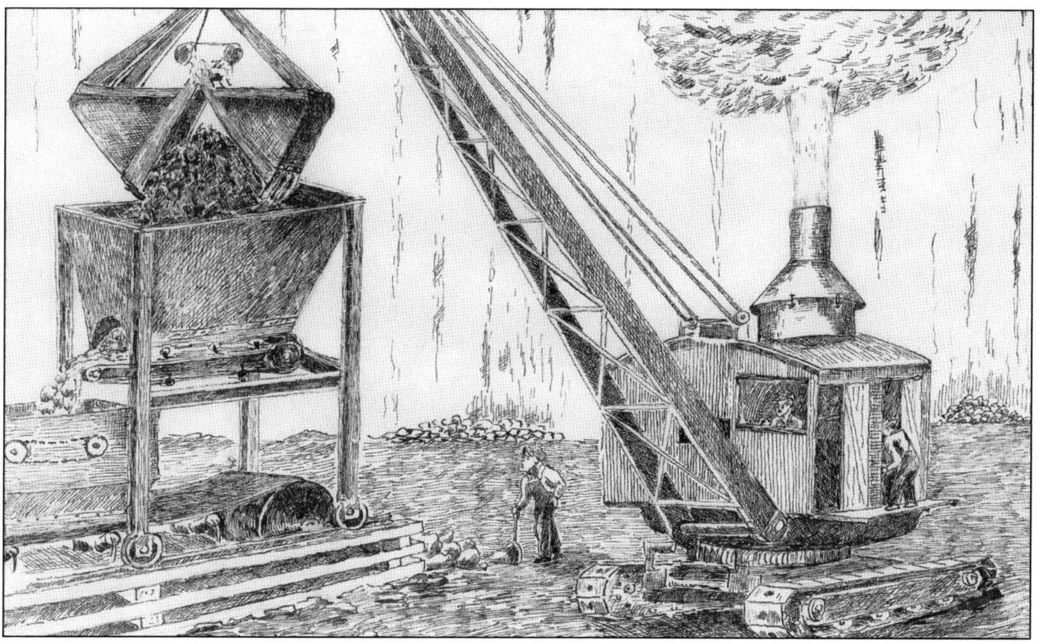

This drawing is titled "Loading Coal with the 'Old Steamer' – 1937." Johnson's preferred art medium was woodburning. He received a woodburning needle for Christmas 1937 and, other than a period when he tried other materials, he was still using a woodburner late in his life. He retired from Wyodak after 30 years and died on September 21, 1996, in Sheridan, Wyoming.

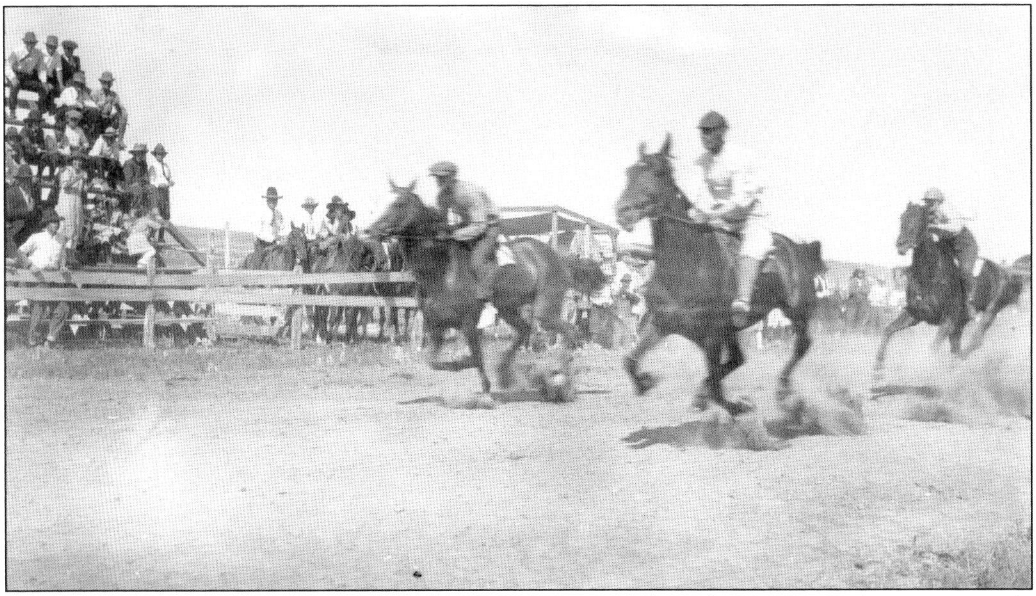

The small town of Wyodak tried to provide as many goods and services as the townspeople needed. The town had its own school, store, and fairgrounds. The fairgrounds were a source of entertainment, hosting such events as horse racing, shown here on July 4, 1925. Spectators had to cross the railroad tracks to get in, and Mr. and Mrs. Preston Gilstrap stationed themselves at the tracks to collect the $1 admission fee.

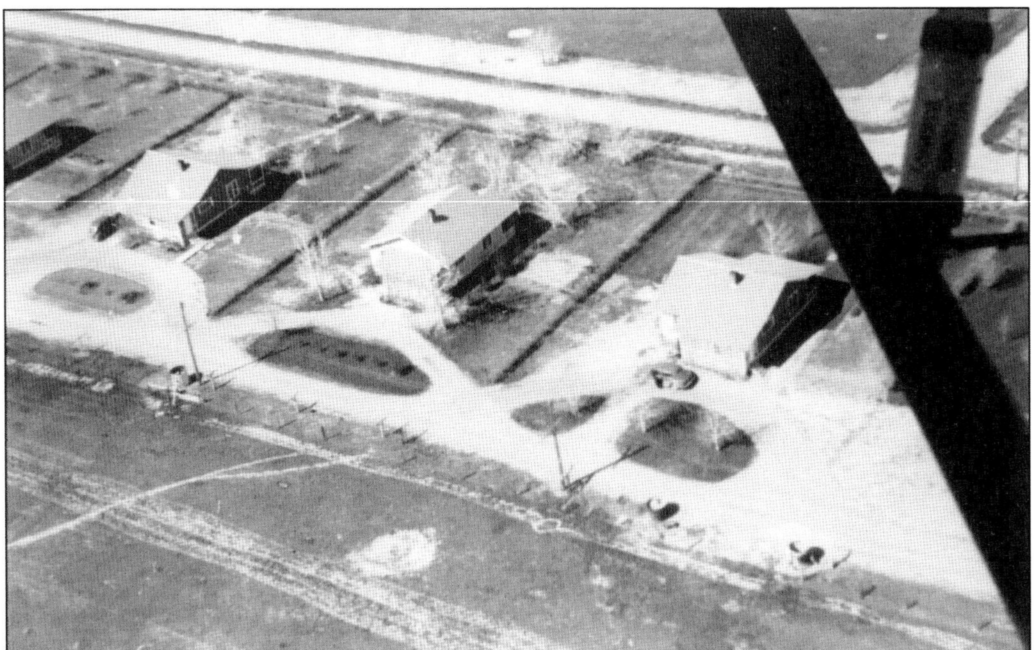

This aerial view shows the houses of Wyodak Village in the 1940s. Residents were told in January 1992 that they had to move by the end of 1995. Black Hills Power & Light Company bought the mine and housing site from Homestake in 1956. By 1992, the company had decided that it did not want to be in the business of running a town.

A piece of Wyodak history died in January 2012, when Ed Ross Jr. passed away at the age of 102. A 1946 mining explosion blinded Ross and left him partially deaf, but he worked despite his disabilities. After working at Homestake, he was transferred to manage the Wyodak Mine, where he stayed for 25 years. Ross's father had started the mine when Ross was in high school, so he was familiar with the operations. (Courtesy of the Black Hills Mining Museum.)

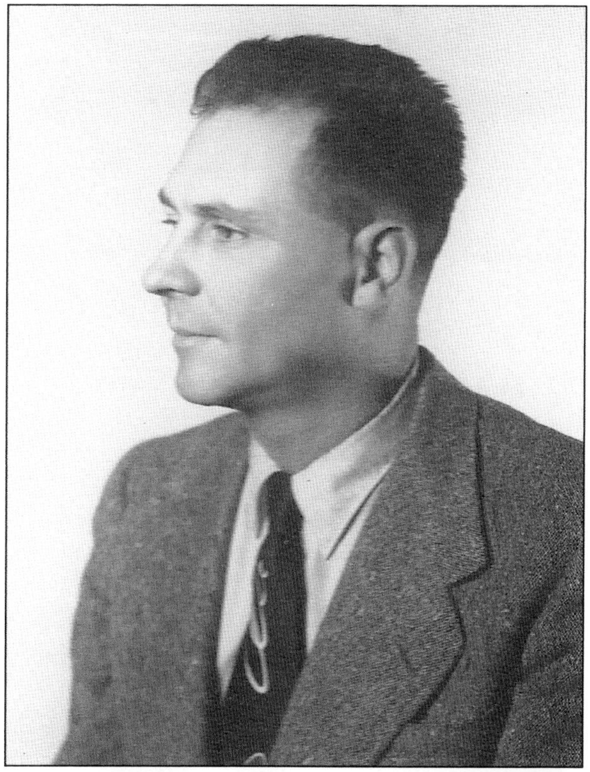

Haul truck capacities have increased tremendously over the years. This early dump truck was most likely used to move overburden (or dirt) when Homestake owned the Wyodak Mine. (Courtesy of Wyodak Mine Archives provided by Kurt Triscori.)

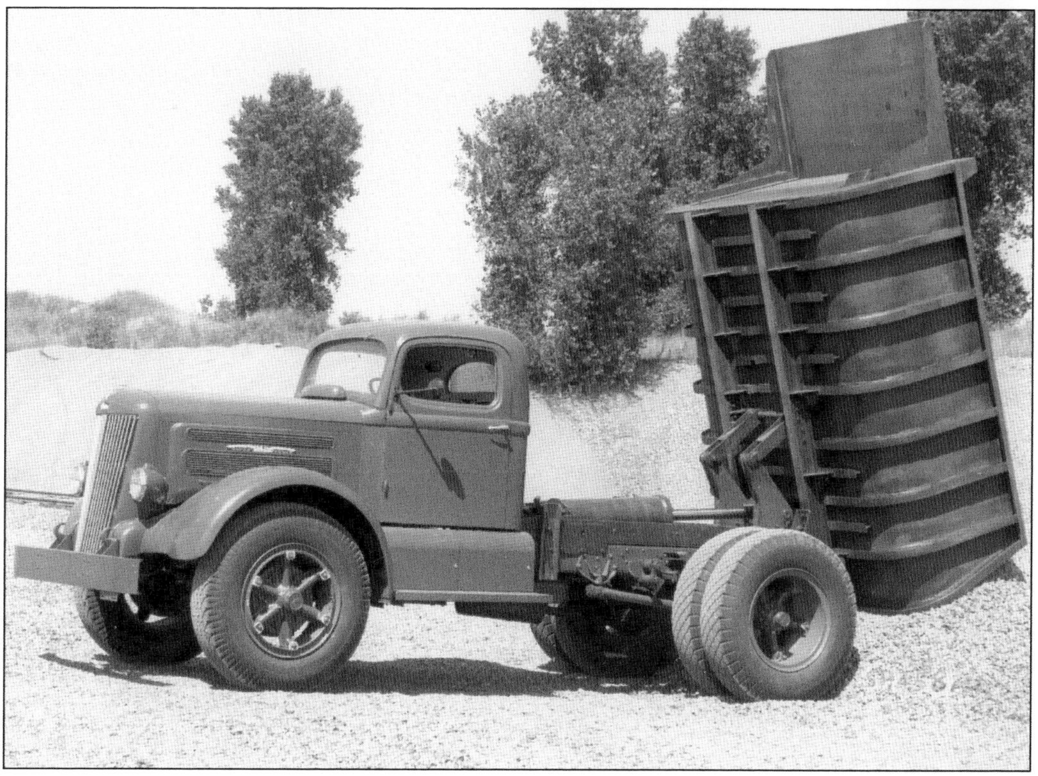

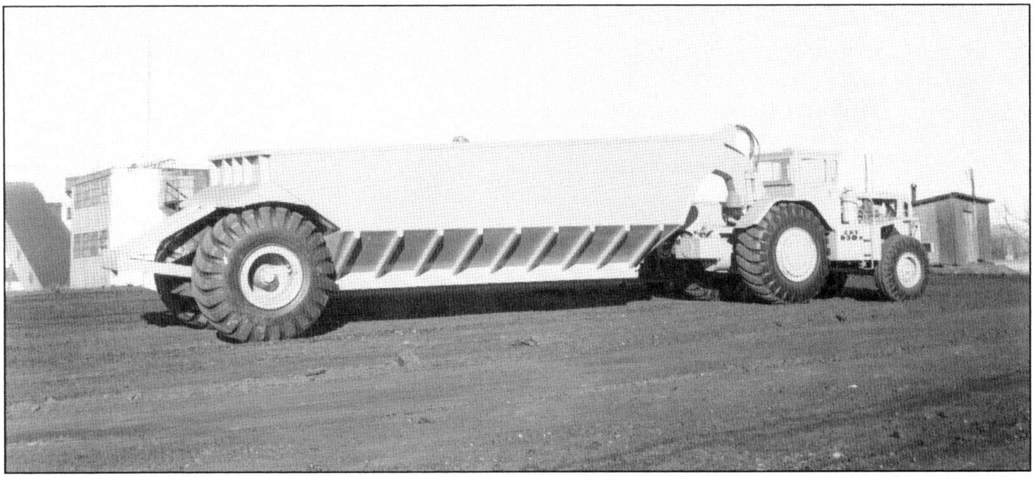

Caterpillar is one of the companies that makes "belly-dump," or "bottom dump," coal haul trucks. While Wyodak used mainly conveyors to move the coal, they did use belly-dumps to transport coal to the Neil Simpson I plant, as well as to the Clovis train load out facility, which was purchased from Kerr McGee. (Courtesy of Wyodak Mine Archives provided by Kurt Triscori.)

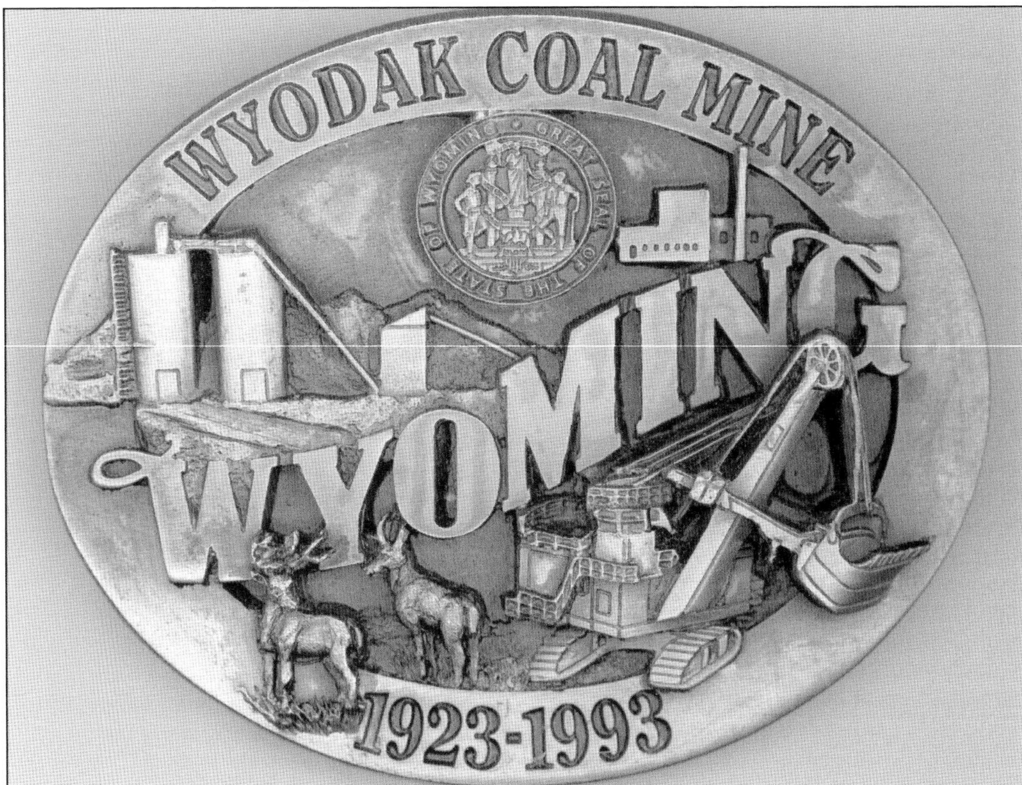

A milestone unmatched in Wyoming is the celebration of 70 years as a continually operating surface coal mine. Belt buckles, documented throughout this book, are popular mementos. Miner's wages have compared well with other industries over the years. In 1975, a dragline operator received an $8.39 average hourly wage. In 2012, the same position paid up to an $36.66 average hourly wage. Wages also vary by region and mine size.

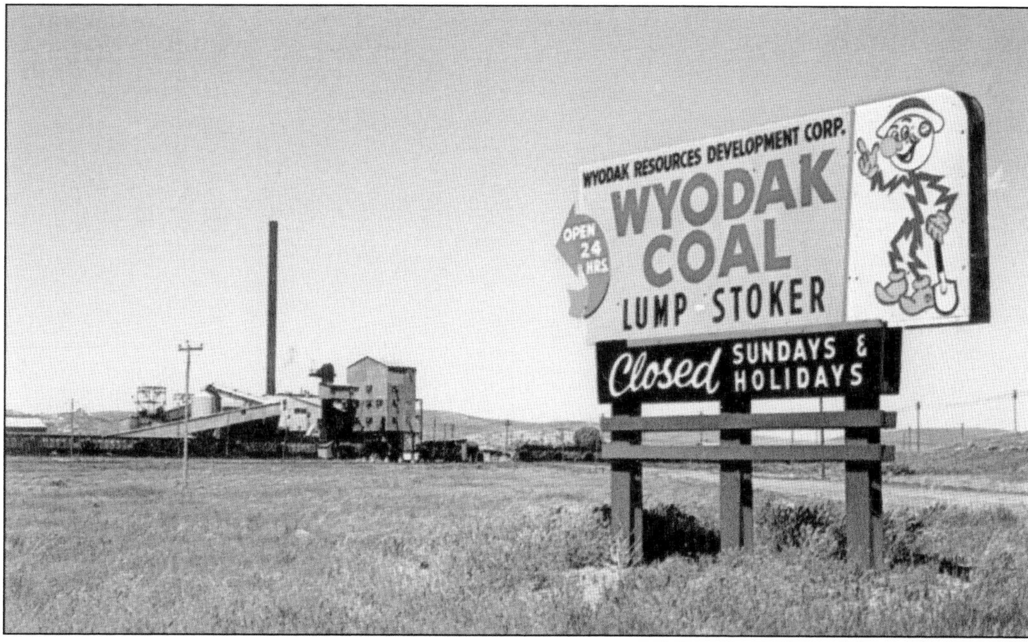

Wyodak was the only mine that sold coal to the general public. Stoker coal is lump coal of a consistent size that can be fed automatically into a coal-burning furnace. Wyodak had two loadout silos (shown below) specifically for coal sales to the public. (Courtesy of Randy Thomas.)

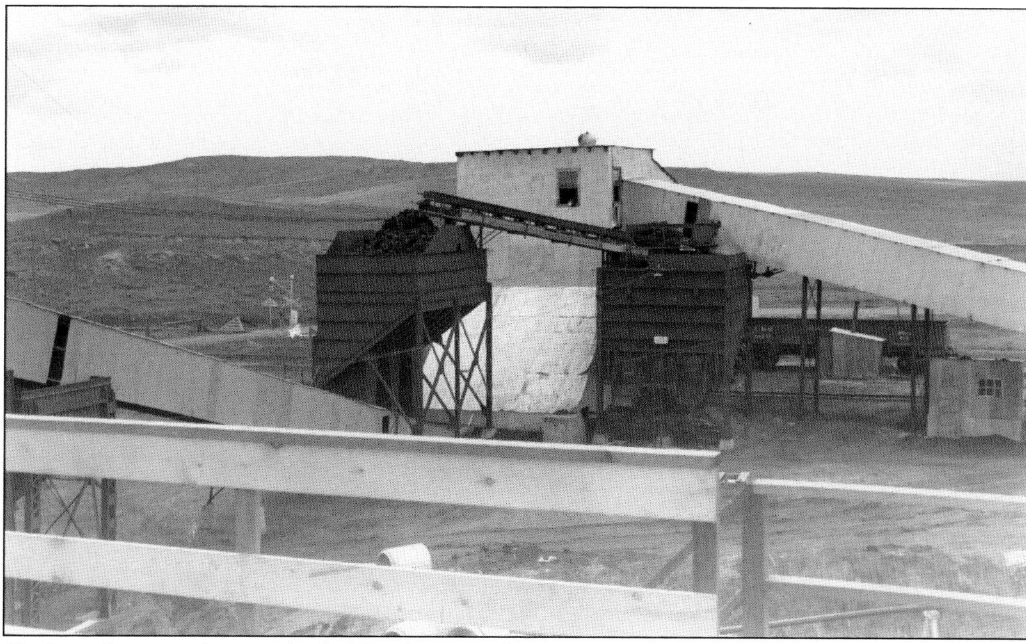
Slack coal is very fine coal—half inch or smaller. During the World War II years in the 1940s, Wyodak used 25 tons of slack coal daily for the operation of its plant and for the production of electrical energy, while the home plant at Lead used 50 tons a day. Every week, 350 tons of slack coal were sent to the West Coast for use in war production plants. This structure is still on Wyodak property today, but is not being used. (Courtesy of Wyodak Mine Archives provided by Kurt Triscori.)

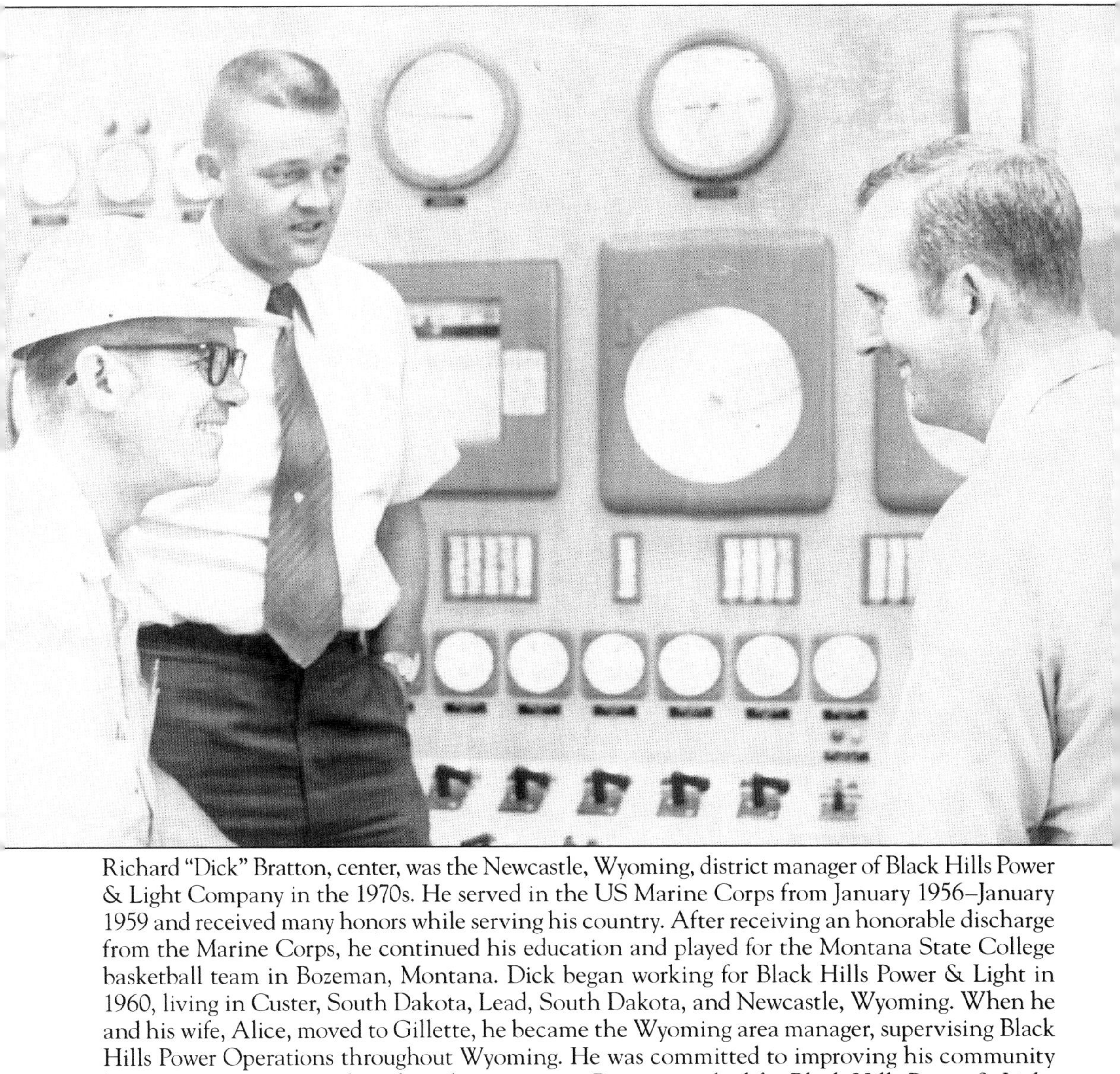

Richard "Dick" Bratton, center, was the Newcastle, Wyoming, district manager of Black Hills Power & Light Company in the 1970s. He served in the US Marine Corps from January 1956–January 1959 and received many honors while serving his country. After receiving an honorable discharge from the Marine Corps, he continued his education and played for the Montana State College basketball team in Bozeman, Montana. Dick began working for Black Hills Power & Light in 1960, living in Custer, South Dakota, Lead, South Dakota, and Newcastle, Wyoming. When he and his wife, Alice, moved to Gillette, he became the Wyoming area manager, supervising Black Hills Power Operations throughout Wyoming. He was committed to improving his community by serving on numerous boards and committees. Bratton worked for Black Hills Power & Light for 44 years. He died on October 14, 2004. The gentleman on the left is Larry Owen, former chief executive officer of Black Hills Power and Light. The gentleman on the right is unidentified.

Seven

Ashland Coal to ARCO to Arch

Ashland, Inc., was founded in 1924 as Ashland Refining Company of Ashland, Kentucky. The company diversified from oil products to paving, construction, chemical distribution, and petroleum. In 1969, the company entered into a joint venture in coal mining under the name Arch Mineral Company. On July 1, 1997, Ashland, Inc., merged two coal companies it controlled, Ashland Coal, Inc. and the Arch Mineral Company, creating the nation's fifth-largest coal producer. Arch Coal, Inc., acquired ARCO's Wyoming coal operations in a joint venture known as Arch Western Resources, LLC, in 1998. The town of Wright was founded in 1976 by Atlantic Richfield (ARCO), which was developing the Black Thunder Mine. The town was named "Wright" after Dale Wright, owner of the Long Branch Bar Ranch, who agreed to sell 735 acres to ARCO to build the town to accommodate the workforce that would be hired to operate the mine. ARCO developed the roads, brought in power and established the water and sewer systems for Wright. Housing Services, Inc., a community development company, was a wholly owned subsidiary of ARCO. They built the enclosed mall, community center, schools, and neighborhoods. On October 15, 1978, Governor Herschler proclaimed "Wright Day" as the official dedication of the town. Thunder Basin Coal Company, another ARCO subsidiary, operates three surface mines in the Powder River Basin—Black Thunder, Coal Creek, and Jacobs Ranch. The story of how Black Thunder got its name begins with Bob Walker, a Texas trail driver who was establishing the T7 Ranch on the Belle Fourche River. According to Walker, he and another rider came upon the winter campground of some Indians who had moved on, but left behind one black pony. When Walker tried to rope the pony, the pony snorted and bucked. Walker called "Whoa there, Black Thunder!" He then led the pony to the T7. Thereafter, the site of the pony's capture was referred to as "the creek where we caught Black Thunder." The creek, and the land basin feeding into it, have since been called Black Thunder and Thunder Basin, respectively.

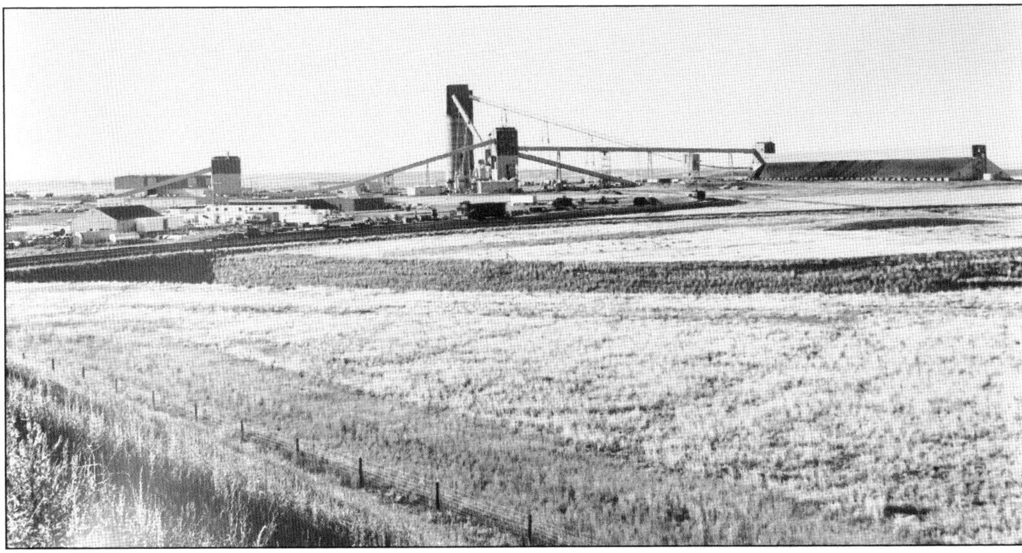

Black Thunder Mine, shown here in the late 1970s, had the distinction of being the nation's largest surface coal mine for years. When Peabody Energy purchased the Rochelle Mine next to their North Antelope Mine creating the North Antelope Rochelle Mine, that complex surpassed Black Thunder in size. However, when Arch purchased Rio Tinto's Jacobs Ranch Mine adjoining Black Thunder, Black Thunder once again became the largest mine in the country. (Courtesy of Arch Coal, Inc.)

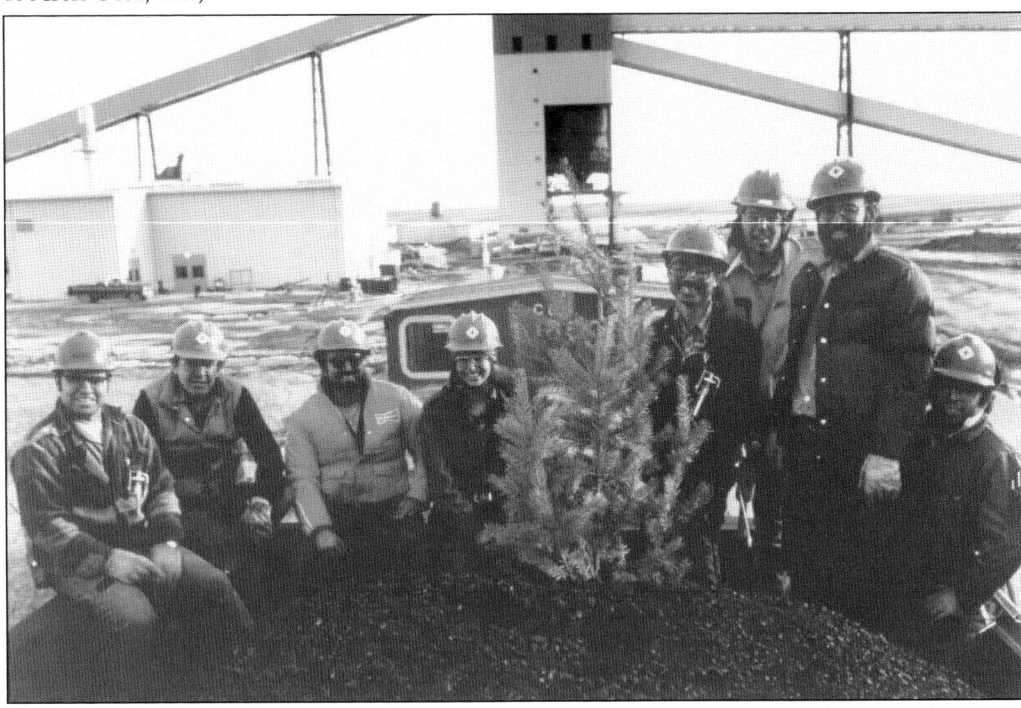

The first train out of Black Thunder occurred on December 14, 1977. An industry tradition was to ship a "good luck tree" in the last car of the first train from a mine. Sitting with the tree are, from left to right, Michael Fielder, Philip Kivisto, John Bradfield, Kay Bock, Mark Duca, Doug Diedrich, Wilson Copeland, and Tom Arreola. (Courtesy of Arch Coal, Inc.)

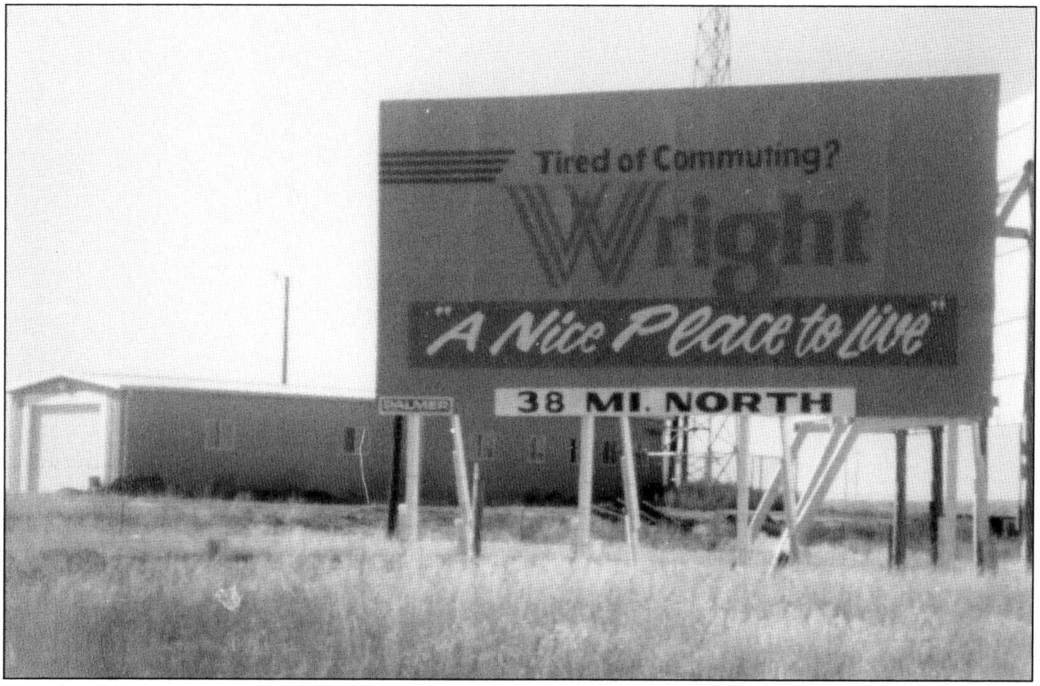

Wright, Wyoming, was founded by Atlantic Richfield as a necessity. At least a dozen surface coal mines had opened during the 1970s, and Gillette was bursting at the seams. When ARCO decided to open the Black Thunder Mine, it also decided to build its own town. Not only would that provide housing for Black Thunder employees, but it would also shorten the 80-mile round trip between Gillette and the mine. (Courtesy of Wright Centennial Museum.)

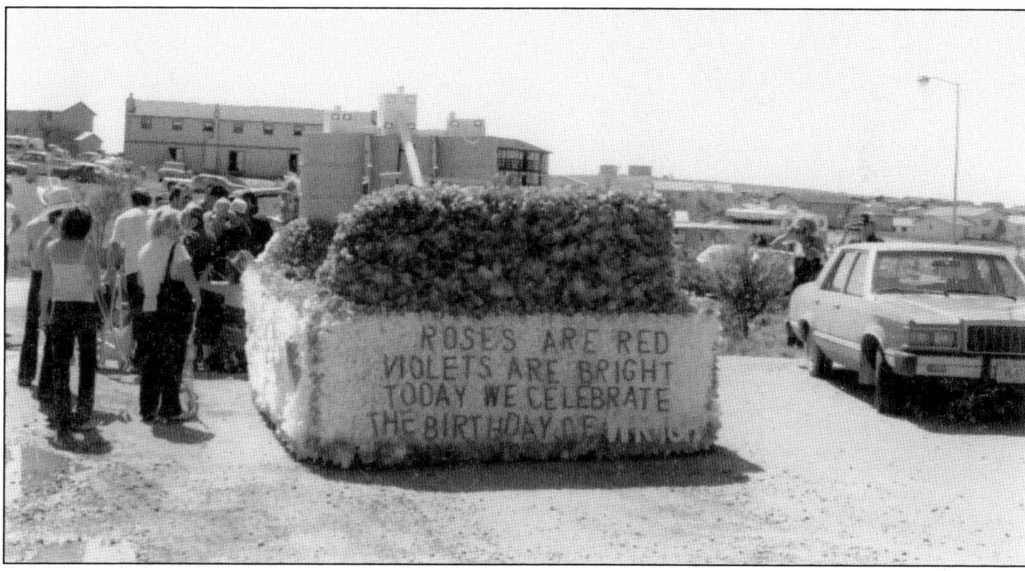

This float celebrates Wright's birthday on October 15, 1978, with the saying, "Roses are red / Violets are bright / Today we celebrate / The birthday of Wright." Atlantic Richfield never wanted to be in the real estate business. They did not establish Wright to make a profit but to help the town get established. The town incorporated in 1985 and elected its own city government. (Courtesy of Wright Centennial Museum.)

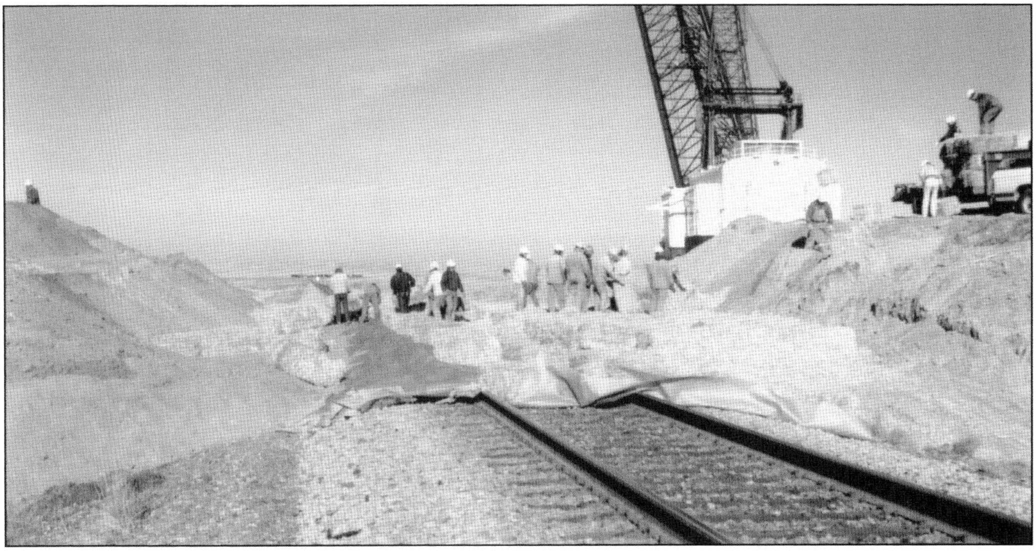

A downturn in the economy idled the Coal Creek Mine in the winter of 1991. The decision was made to move the Bucyrus Erie 1300W dragline from Coal Creek Mine to Black Thunder Mine, where it could help increase production. The dragline, named "Walking Stick," weighed 4.5 million pounds. Preparations had to be made for the dragline to walk the Wyoming prairie. (Courtesy of Arch Coal, Inc.)

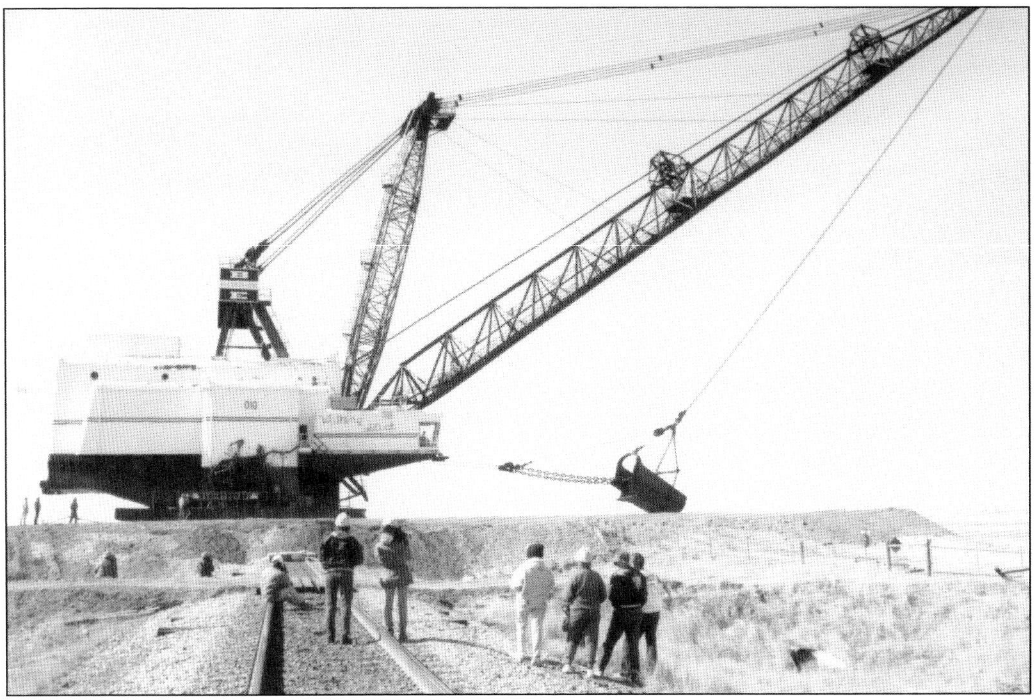

Dirt and straw were spread over the railroad tracks so the dragline could cross more easily. The long truss-like boom is 302 feet long, and controls a bucket that holds 45 yards of overburden. Draglines move overburden, not coal, and run on electricity. Huge electric cables connect to nearby power sources or mobile generators. Draglines are typically built at the mine site and can take years to assemble. (Courtesy of Arch Coal, Inc.)

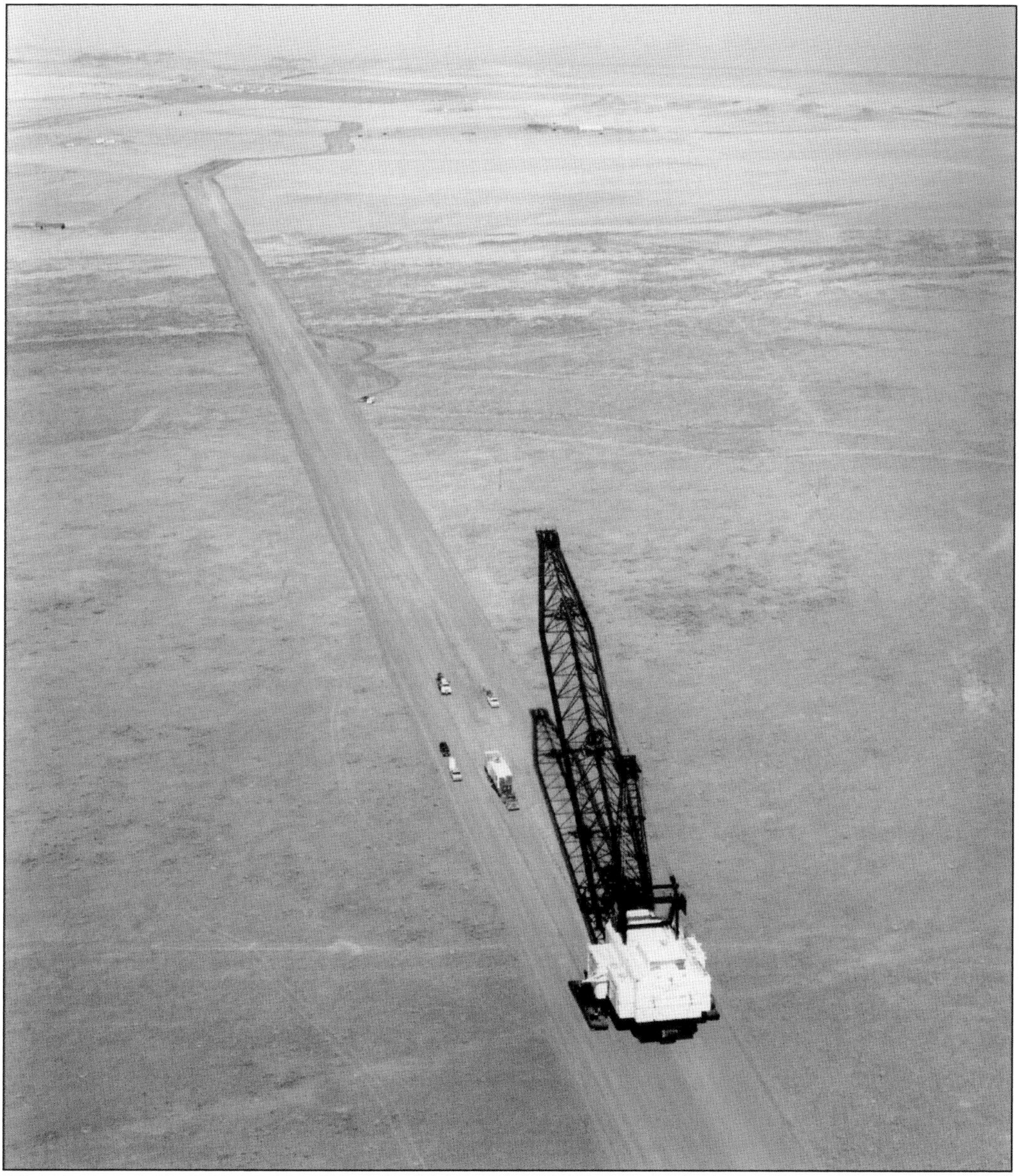

The distance from Coal Creek to Black Thunder is 26 miles, and it took 19 days to travel the distance. Draglines travel approximately three feet per minute. Its walking motion is accomplished by lowering the float-like "shoes" on each side of the dragline and lifting the back. As the shoes go through a 60-second cycle, they drag the entire machine backwards seven feet and then lower it to the full down position on the ground. Despite their massive size, modern draglines are easy to operate using joystick controllers and computer-assisted drives. The operator has an ergonomically adjustable seat for comfort during a 12-hour shift. Two operators run the dragline. One sits in the cab at the control station, and the other makes routine maintenance checks inside the machine. They switch positions every hour to break the monotony and stay alert. Black Thunder has six draglines currently, more than any mine in the Powder River Basin. (Courtesy of Arch Coal, Inc.)

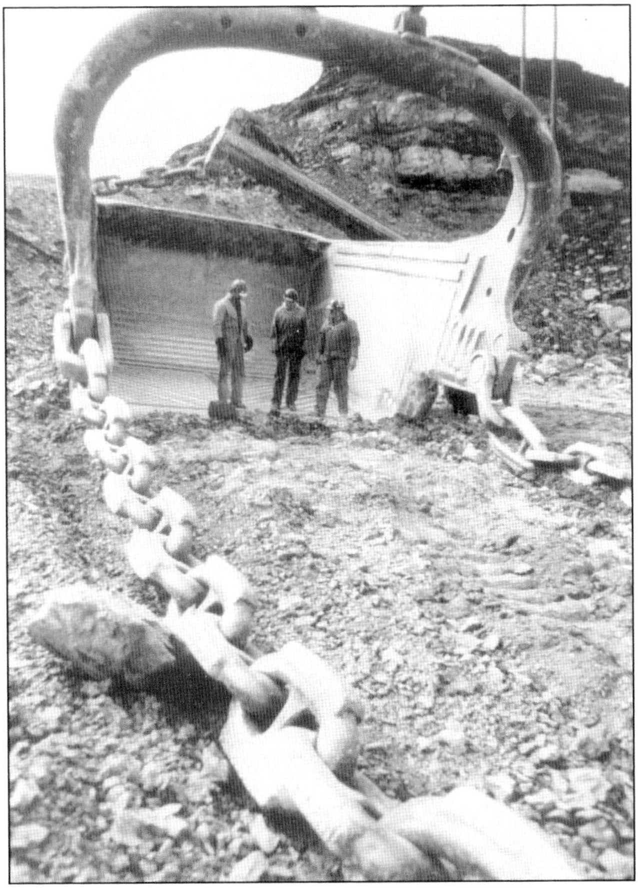

The bucket of the Bucyrus Erie 1570W dwarfs the three men standing inside. When a new piece of equipment is acquired by Arch, the company holds a contest for the employees to name the piece. This bucket belongs to the dragline named Thor, for the mythical god of thunder. (Courtesy of Arch Coal, Inc.)

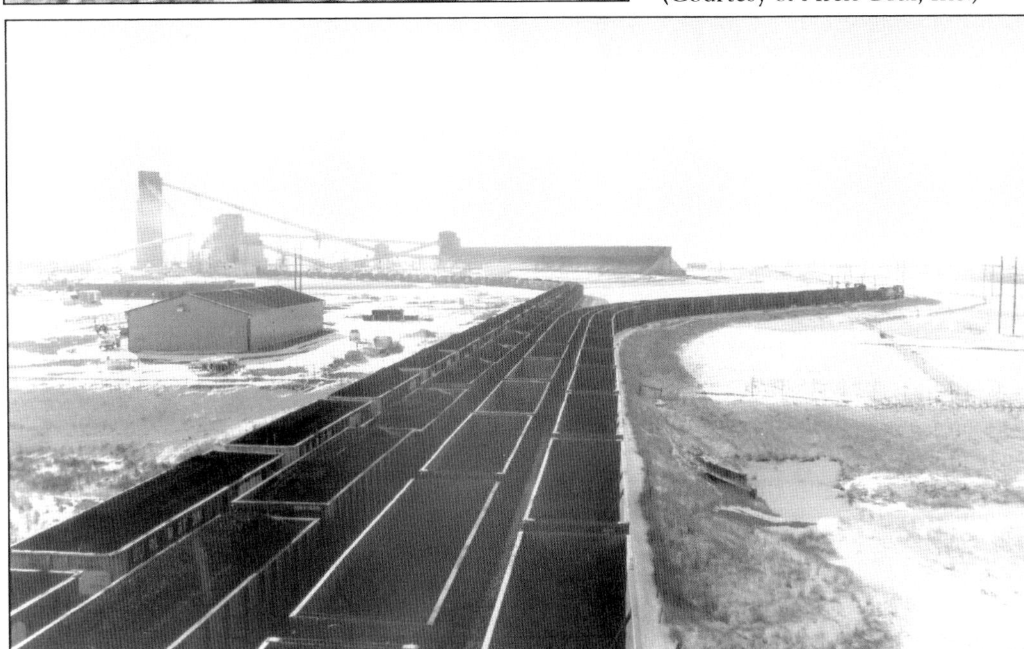

A snowy scene from December 1990 shows Black Thunder's four sets of railroad tracks where coal trains line up, waiting to be loaded. The railroad originally shipped cattle and, for a time, Gillette shipped more cattle than any other shipping point on any other railroad in the world, according to Wyoming author Bill Bragg. (Courtesy of Arch Coal, Inc.)

Larry Curtis started working at Black Thunder in 1981. He was in the blasting department in 1992 when the Board of Directors of Arch came to visit the mine. Larry has millisecond, or "ms," connectors on his arm. The coal had been drilled with holes to hold explosives, and Larry was in charge of loading the holes. (Courtesy of Arch Coal, Inc.)

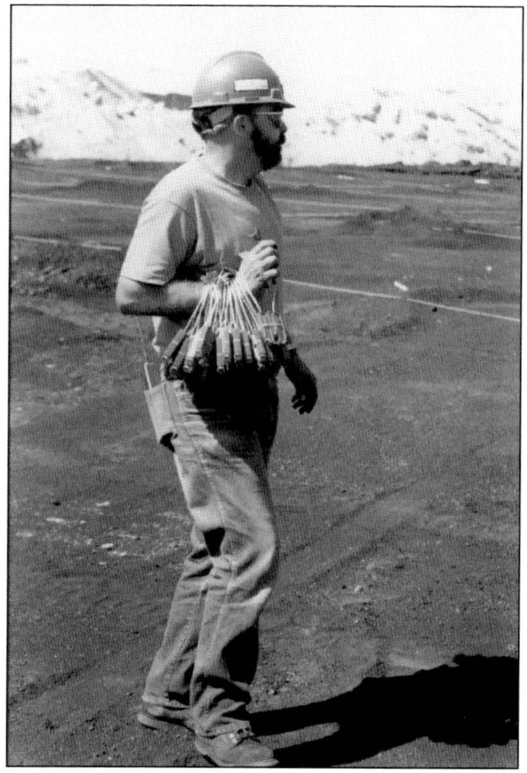

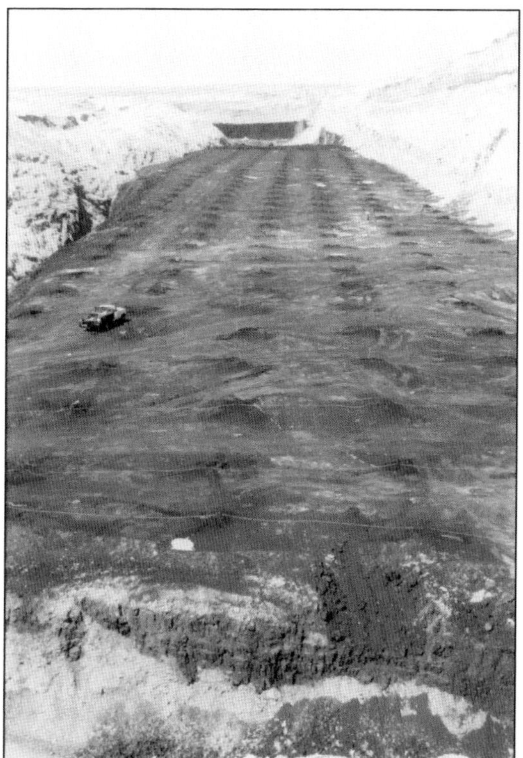

Blasting can be used to loosen both coal and overburden. Drilling rigs drill holes up to approximately 75 feet deep. (Courtesy of Arch Coal, Inc.)

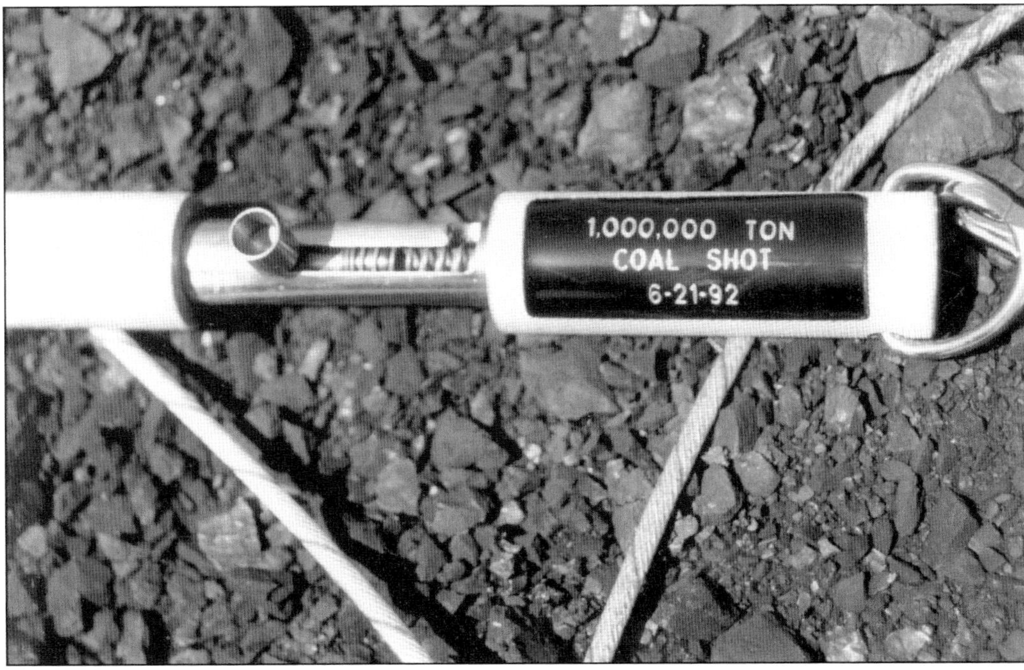

As a special demonstration for the board of directors, one million tons of coal were shot at the mine, imprinted on the shooter shown here. The date was June 21, 1992. Electronic blasting began being used around 2004. An E-cord burns at the rate of 22,000 to 25,000 feet per second to detonate the explosive. (Courtesy of Arch Coal, Inc.)

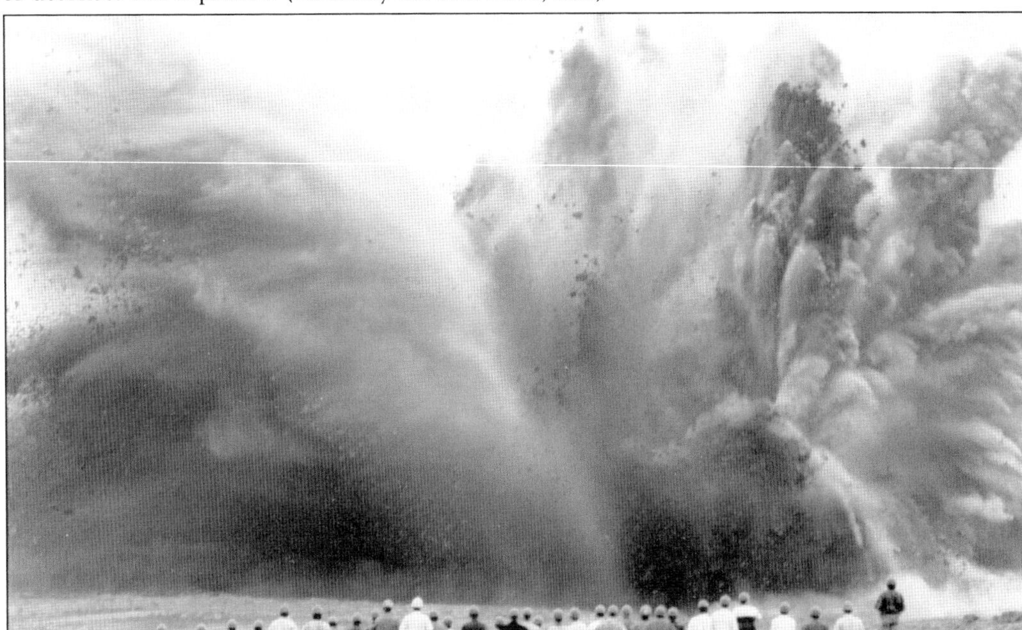

This blast at Black Thunder had an audience of about 34 visitors. As exciting as the blast appears, it does not happen like this anymore. Fly rock and overburden are more contained with the modern-day blast. A cast blast is designed to use the explosive energy to move overburden material across the pit. This is of major cost savings to the mine. (Courtesy of Arch Coal, Inc.)

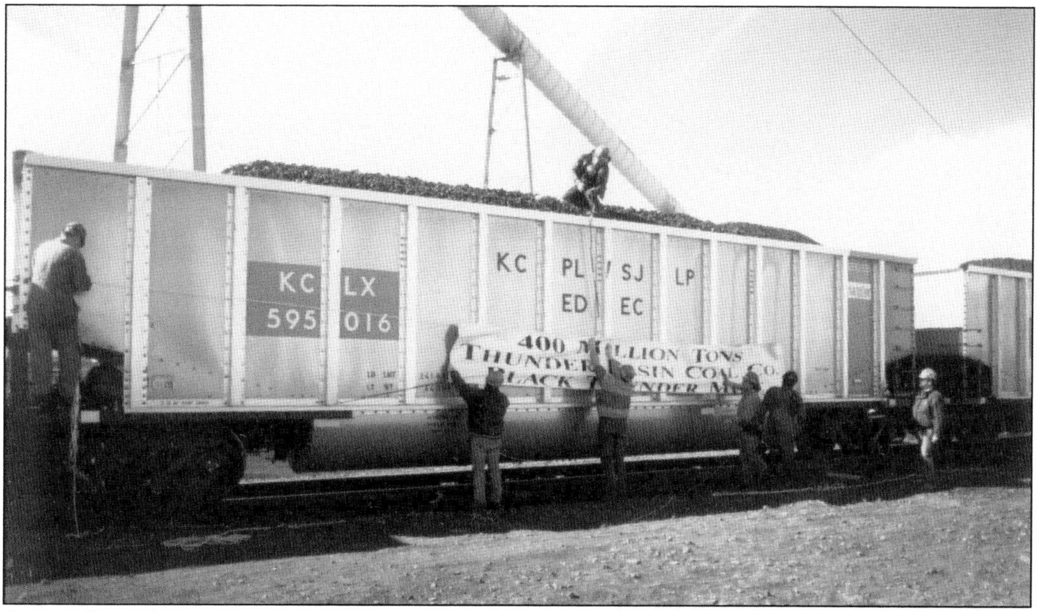

A train bound for Kansas City Power & Light is decorated with a banner to celebrate 400 million tons. Black Thunder's fleet of haul trucks includes Liebherr and Komatsu trucks. Coal is hauled from the pit to a dump and crusher station near the pit, which feeds an overland conveyor to the coal storage and loadout silos. Both the Burlington Northern and Union Pacific serve the mine. (Courtesy of Arch Coal, Inc.)

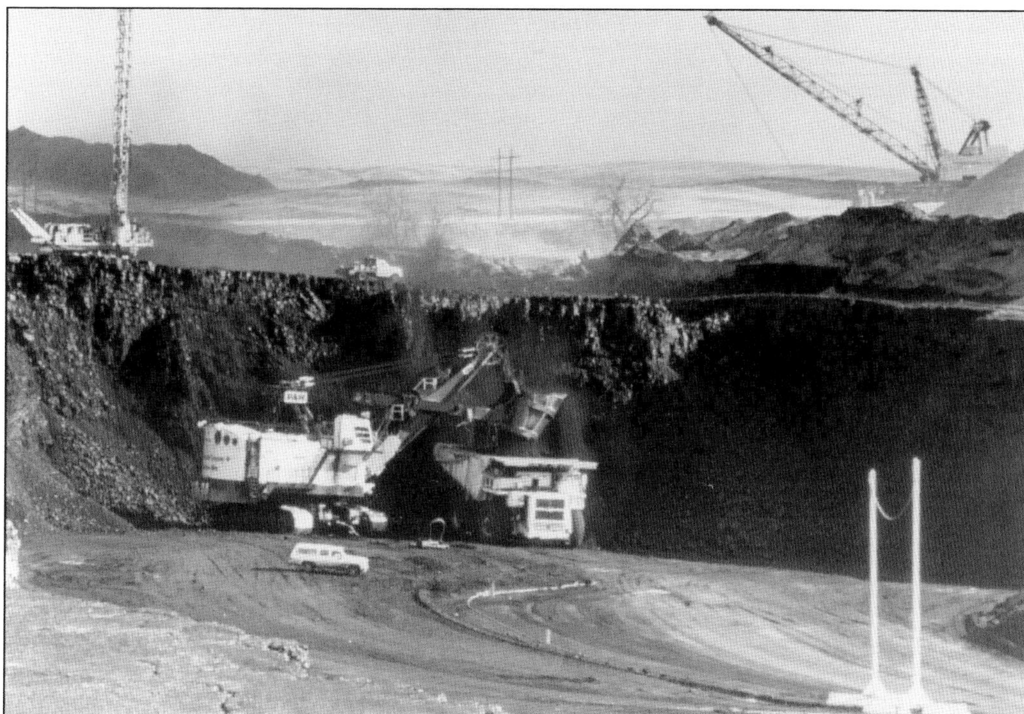
All the significant large equipment is shown here in the mine pit—upper left is a drill, upper right is a dragline, lower left is a shovel, lower right is a haul truck. (Courtesy of Arch Coal, Inc.)

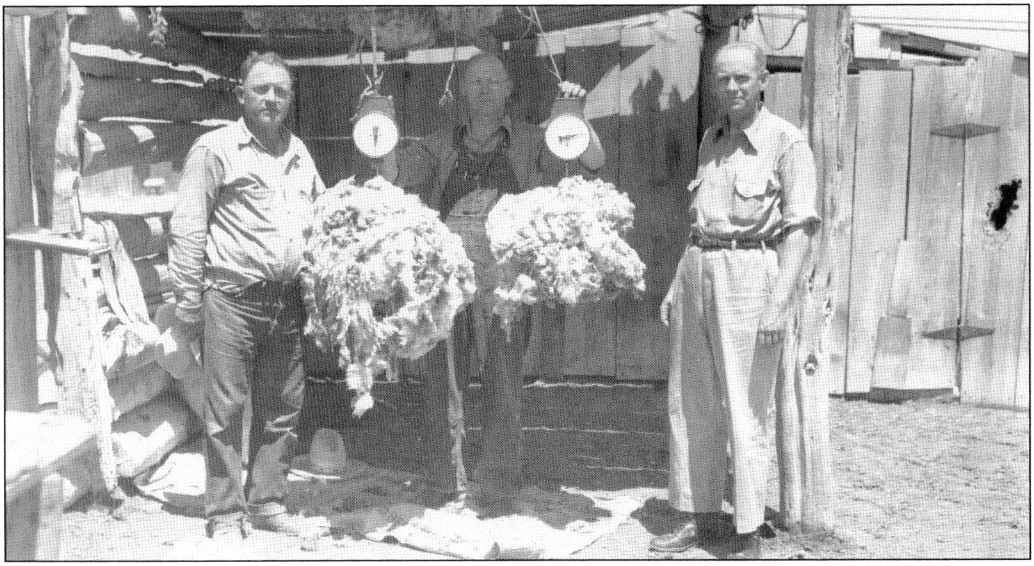

John A. "Gus" Jacobs, left, was born in Kentucky in 1902. He ran a sheep ranch 15 miles southeast of where the town of Wright would later be located. Gus sold his sheep ranch to Kerr McGee Coal Corporation in the 1970s, when many coal mines were run by oil companies. By 1977, Jacobs Ranch Mine was producing coal.

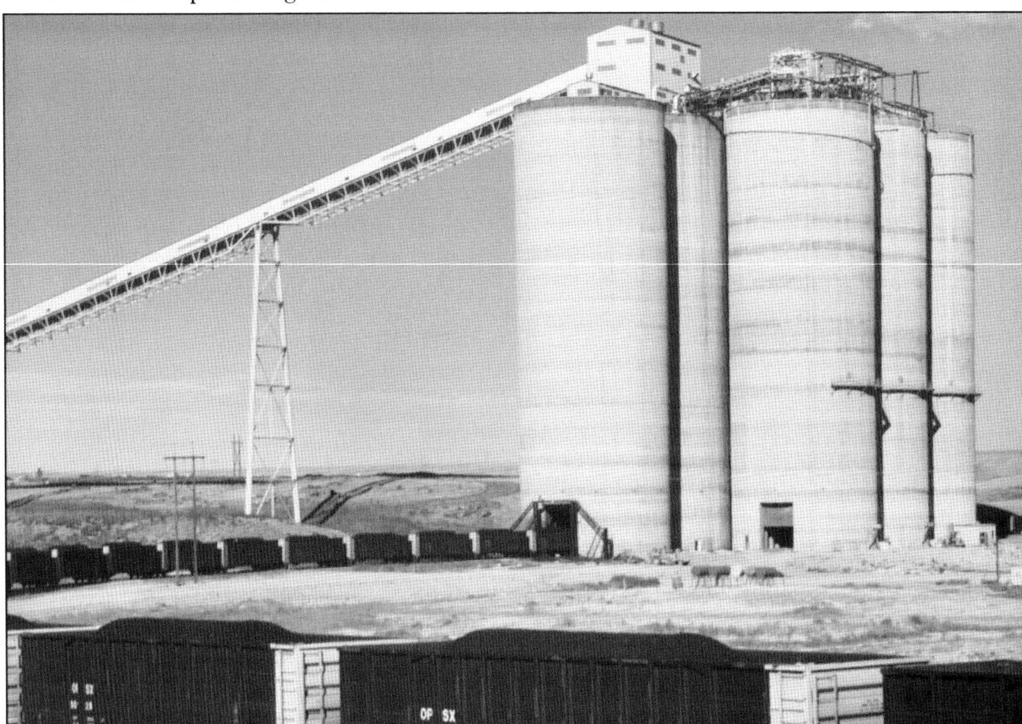

Jacobs Ranch Mine occupies 15,000 acres and is located across Wyoming Highway 450 from Black Thunder Mine. Kerr McGee sold Jacobs Ranch to Rio Tinto Energy America in 1998. In early 2009, Rio Tinto sold Jacobs Ranch Mine to Arch Coal for $761 million. The silos of Jacobs Ranch Mine are shown here in 1980.

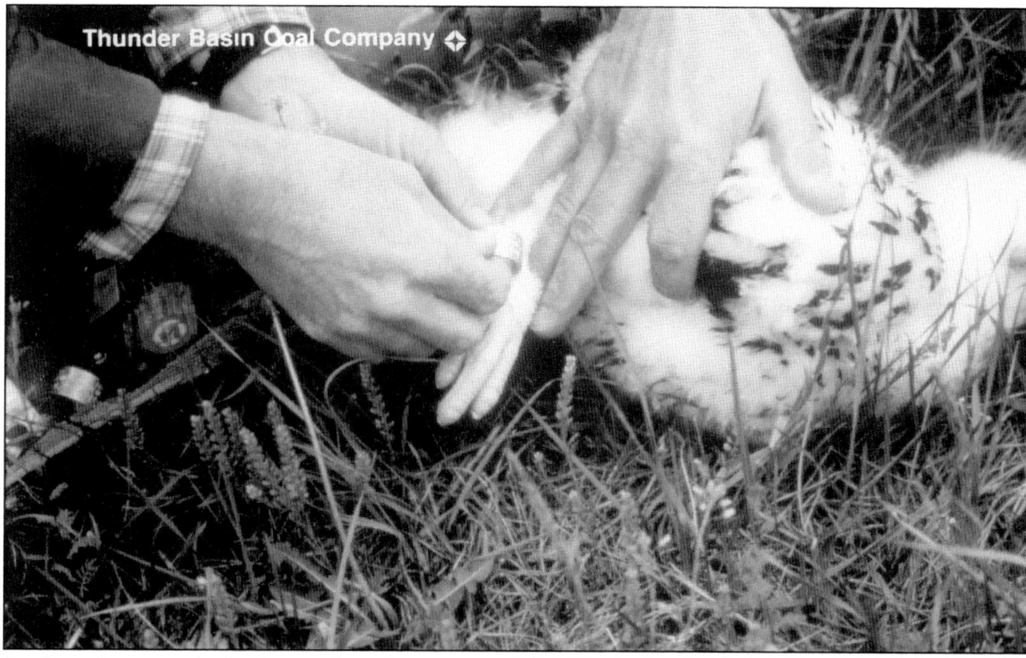

In this photograph, biologists band a young golden eagle on Coal Creek property. They then recorded the serial number on the band with the US Fish and Wildlife Service. Golden eagles are one of the most successful birds of prey in North America, due in part to their protection under two federal laws, which also protect the bald eagle. (Courtesy of Arch Coal, Inc.)

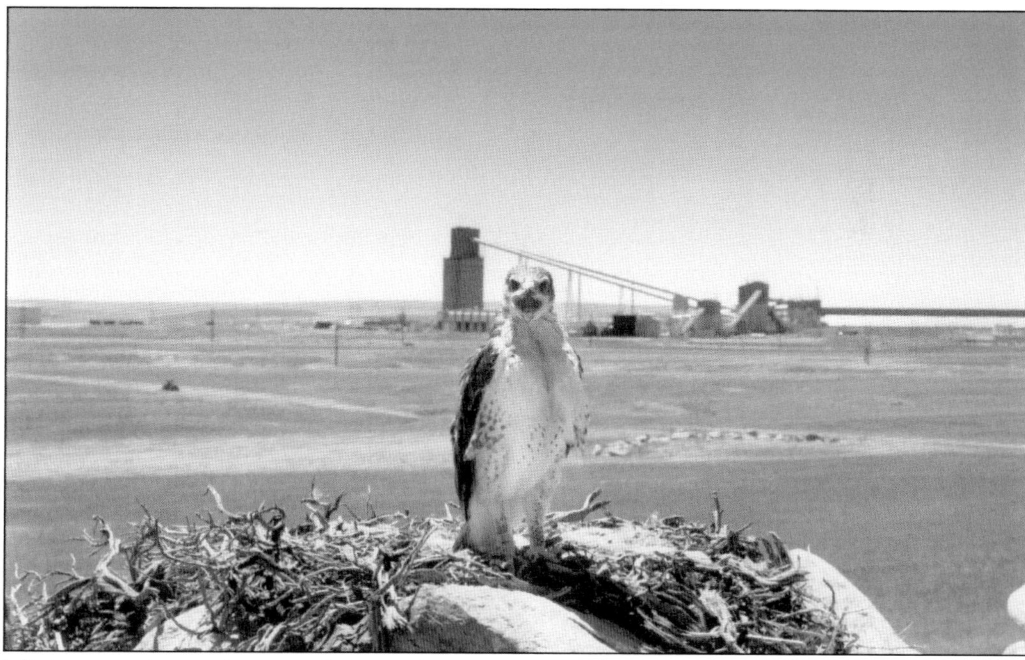

Mining permits require severely limited activities around nests, unless the birds can be enticed to move their nests. In 1979, ARCO studied whether birds would use a man-made structure instead of a tree for their nests, and were successful in moving the birds safely out of the way until mining was finished. (Courtesy of Arch Coal, Inc.)

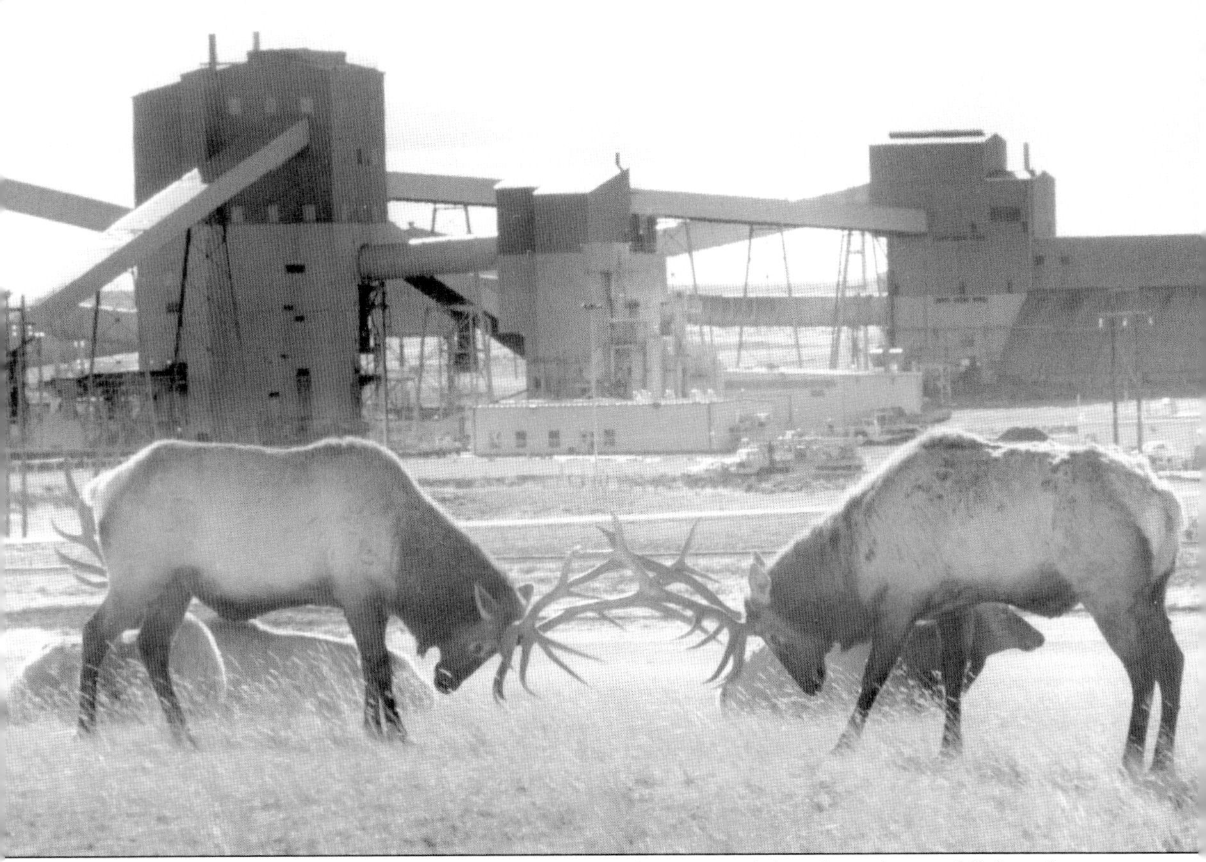

This iconic photograph has been made into a popular postcard that shows how wildlife and mining can exist side by side. Not all mines in the Powder River Basin can boast of an elk population, but those mines in the southern end of the county, like Black Thunder, are lucky to have an elk herd as residents. The elk, or wapiti, is one of the largest species of deer in the world, and one of the largest land mammals in North America. In North America, males are called bulls and females are called cows. Male elk have large antlers, which are shed each winter. During the mating period, mature bulls compete for the attention of the cows. Rival bulls challenge opponents by bellowing and by paralleling each other, walking back and forth. This allows potential combatants to assess the other's antlers, body size, and fighting prowess. If neither bull backs down, they engage in antler wrestling. Elk are a ruminant species with a four-chambered stomach, feeding on grasses, plants, leaves, and bark. (Courtesy of Arch Coal, Inc.)

Eight

Buckskin and Dry Fork

Dry Fork Station, owned by Basin Electric Power Cooperative and the Wyoming Municipal Power Agency, is the newest coal-based power plant opened in Gillette, Wyoming. According to some organizations, it could be among the last coal-based power plant to be permitted in the United States. The plant was dedicated August 11, 2011, with more than 1,100 people in attendance, including Gov. Matt Mead; former governor Dave Freudenthal; Sen. John Barrasso; Rep. Cynthia Lummis; Ron Harper, Basin Electric CEO and general manager; Cliff Gjellstad, president of Basin Electric's board of directors; Mike Easley, CEO of Powder River Energy Corporation; Steve Hughes, chairman of the Campbell County Commission; and Tom Murphy, mayor of Gillette. Gov. Matt Mead said, "Coal is part of our past, present and future." Western Fuels Association manages the Dry Fork Mine, which is owned by Western Fuels-Wyoming. Dry Fork Mine began producing coal in 1990 and is known as the mine with the least overburden of all Gillette mines. Dry Fork refers to a fork of the Little Powder River that runs through Campbell County. The other mine in this chapter is Buckskin, currently owned by Kiewit Mining Group. Buckskin was developed by Shell Oil Company on a federal lease that Shell purchased from the Farmer's Union Central Exchange in 1975. Buckskin has shipped more than 406 million tons of high-quality, low-sulfur coal since production began in 1981. The use of multiple silos, in-pit crushing, and an on-line quality analyzer allows crews to blend a "super compliant" fuel that meets customers' exact specifications. The Kiewit Corporation was founded in 1884 by Peter Kiewit, a bricklayer of Dutch descent. The founder's son, also named Peter Kiewit, led the company from 1924 until his death in 1979. During his lifetime, it became one of the largest construction companies in the world. The company diversified into engineering, mining, offshore services, and several other subsidiaries. In 2004, Triton Coal Company owned Buckskin Mine. In a complicated transaction that same year, Kiewit became the owner of Buckskin. Its sister mine, North Rochelle, became the property of Arch Coal, Inc.

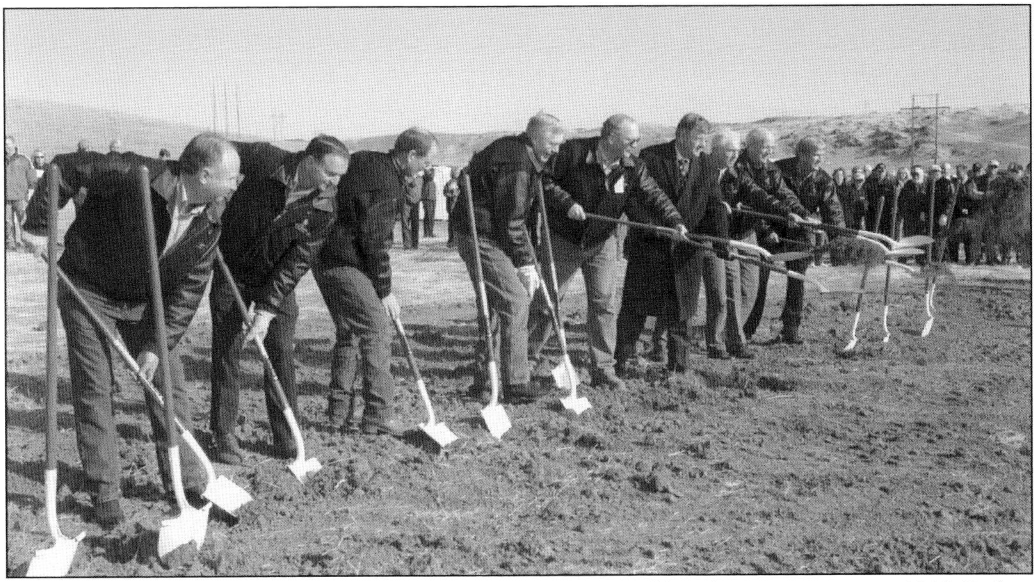

Construction on the Dry Fork Station began with this ground-breaking ceremony in 2007. Before construction could begin, Basin Electric had to obtain several regulatory permits, including an Industrial Siting Permit and an air permit. On October 17, 2007, two days after the air permit from the State of Wyoming was in hand, construction on the power plant began. The plant cost $1.35 billion to build. (Courtesy of Basin Electric Power Cooperative.)

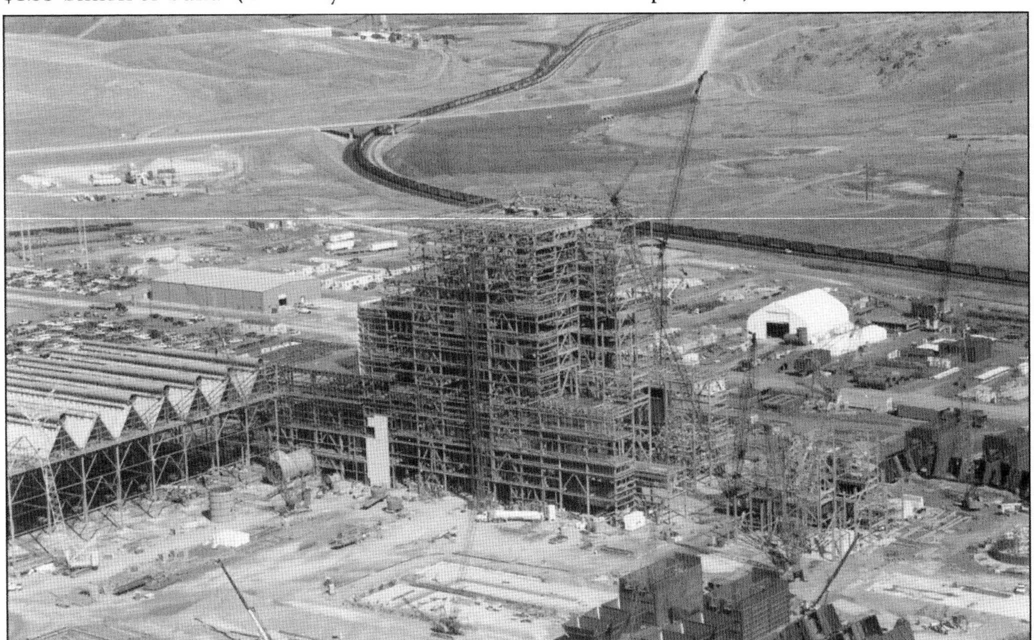

During peak construction in 2009, more than 1,300 workers from over 36 states were working on the plant. The workforce accumulated more than six million man-hours without a lost-time incident during construction. Basin Electric Power Cooperative is a consumer-owned regional cooperative headquartered in Bismarck, North Dakota. It generates and transmits electricity to 135 member rural electric systems in nine states. (Courtesy of Basin Electric Power Cooperative.)

The Dry Fork Mine produces approximately six million tons of coal annually. Customers include Basin Electric's Laramie River Station in Wheatland, Wyoming, the Leland Olds Station in Stanton, North Dakota, and the newly constructed Dry Fork Station. This photograph shows construction of silos No. 4 and No. 5 around June 2009. The additional silos were built to support the new Dry Fork Station. (Courtesy of Basin Electric Power Cooperative.)

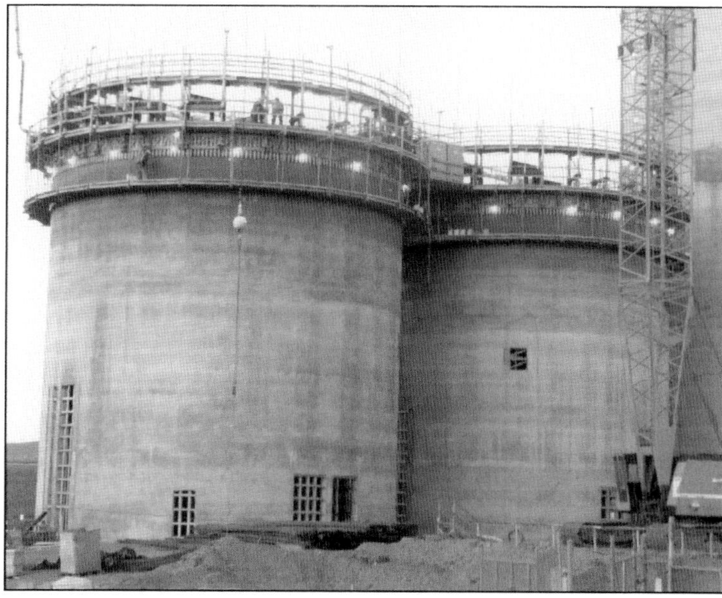

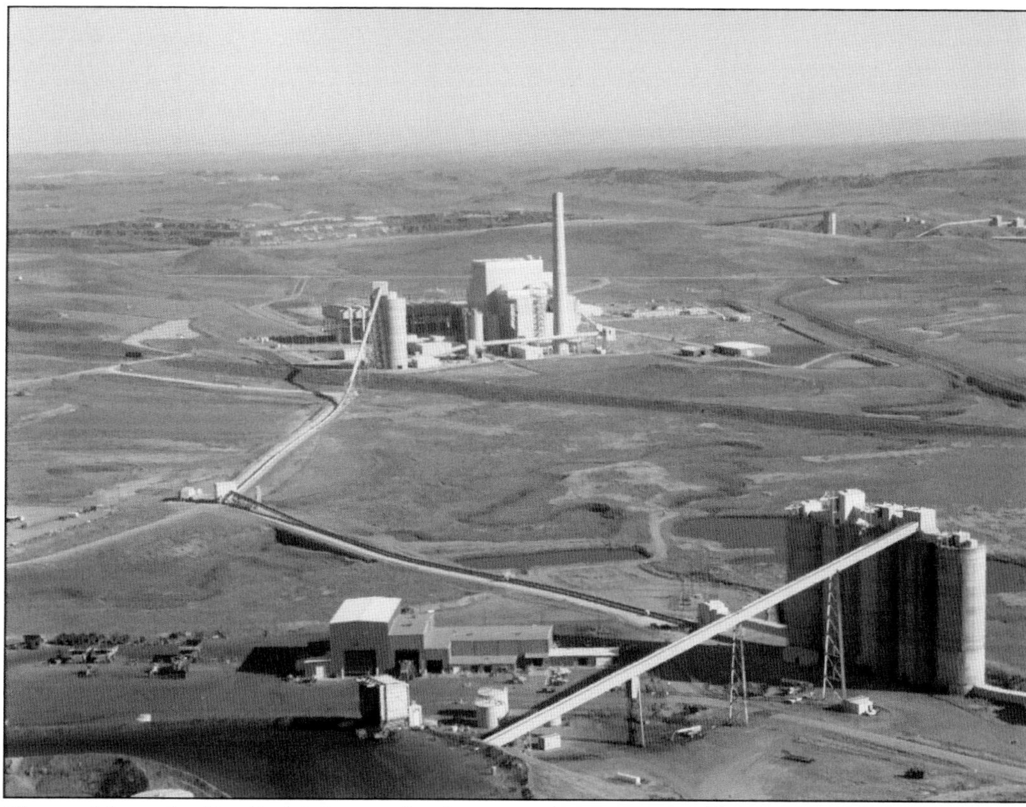

The coal from the Dry Fork Mine in the foreground is transported by a one-mile conveyor to the Dry Fork Station in the background. In 2010, the Dry Fork Mine won a prestigious Excellence in Surface Coal Mining Award from the US Department of the Interior for exemplary compliance during construction of a wetlands crossing project that was significantly more extensive than regulations required. (Courtesy of Basin Electric Power Cooperative.)

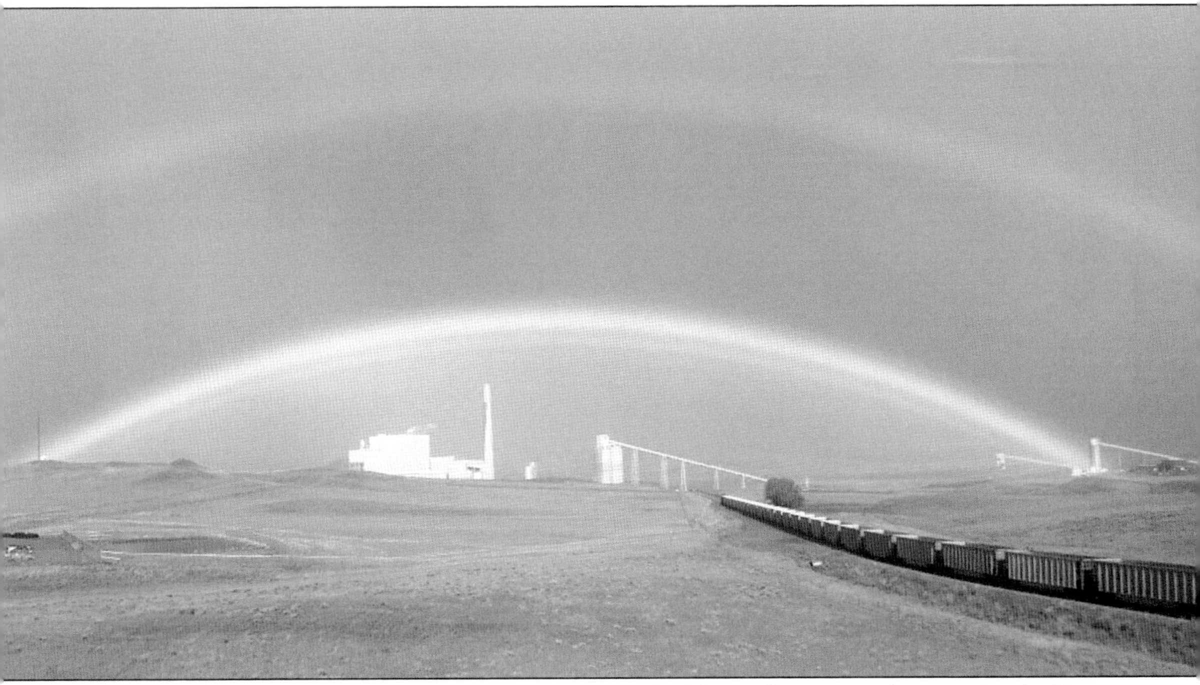

This photograph of the Dry Fork Station under a double rainbow was taken in 2011 by Dennis Thorfinnson, the Dry Fork Station plant engineer. The photograph so impressed the Basin Electric management team, it has become a signature photograph for the power plant. (Courtesy of Dennis Thorfinnson.)

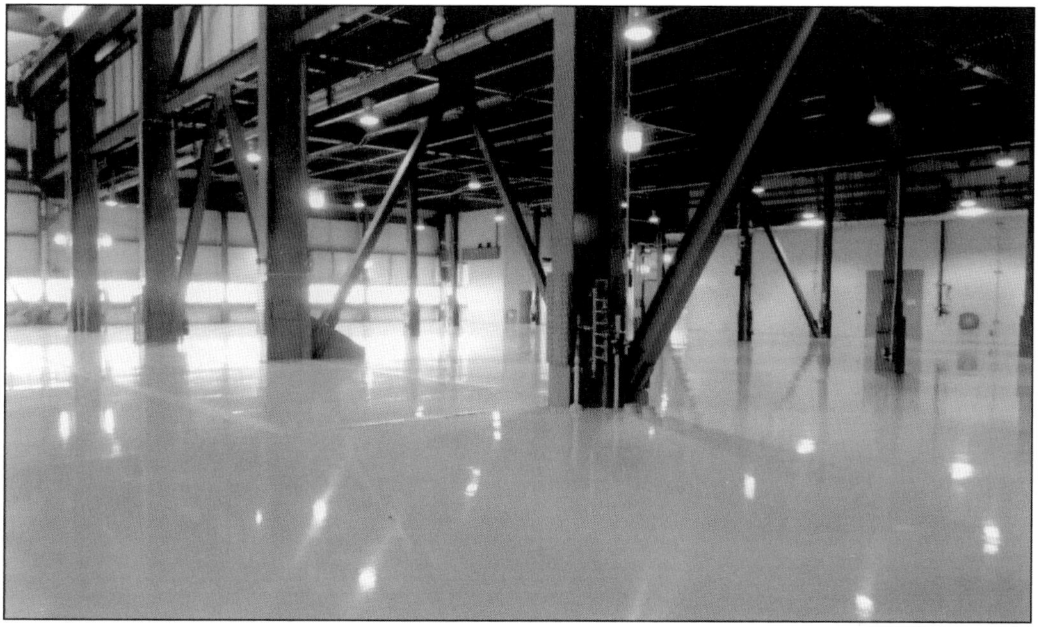

Most people would not think a coal-fired power plant would shine with cleanliness but this one does. More than $336 million has been invested at the Dry Fork Station for advanced environmental technologies for capturing and removing emissions of sulfur dioxide, mercury, ash, and oxides of nitrogen. (Courtesy of Basin Electric Power Cooperative.)

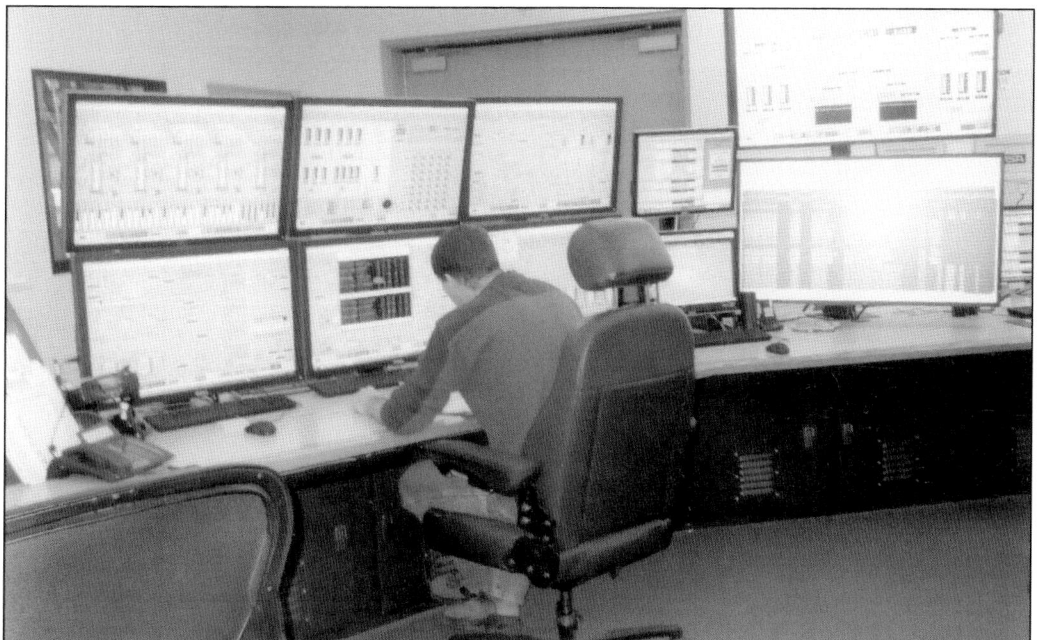

One of the amazing facts about a contemporary coal-fired power plant like the Dry Fork Station is the minimal number of employees it takes to run a facility of this size. Only 85 employees staff the Dry Fork Station, including Wayne Knudsen, lead station operator, pictured here. With his series of computer screens, he can oversee and maintain operations at the plant. (Courtesy of Basin Electric Power Cooperative.)

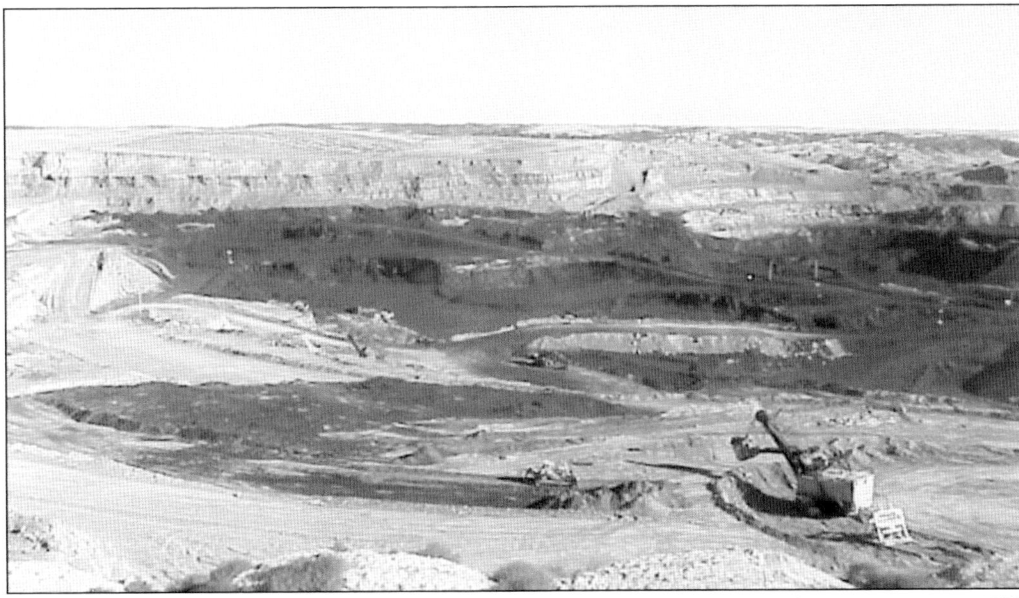

Buckskin Mine, shown here in 2000, was the site of an archeological excavation in 1977, approximately five years before production. The excavation revealed bison bones, a camel bone, projectile points, and butchering tools dating from around 9,000 years ago. George C. Frison, head of the University of Wyoming Archaeology Department, led the summer project. Both Carter Mining and Kerr McGee contributed $34,000 for the study. (Courtesy of Buckskin Mine Employees.)

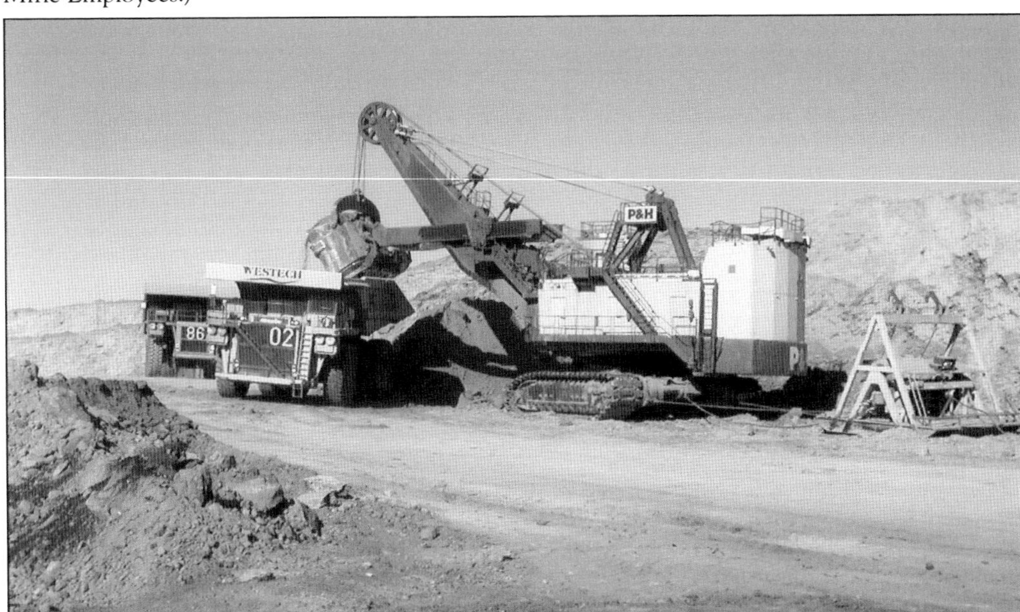

P&H MinePro is a supplier of major mining equipment, such as blasthole drills, draglines, electric shovels, mobile crushers, wheel loaders, and conveyors. P&H is the support and distribution arm of Joy Global Surface Mining. This P&H 2800 shovel is loading a 240-ton haul truck at Buckskin Mine. (Courtesy of Buckskin Mine Employees.)

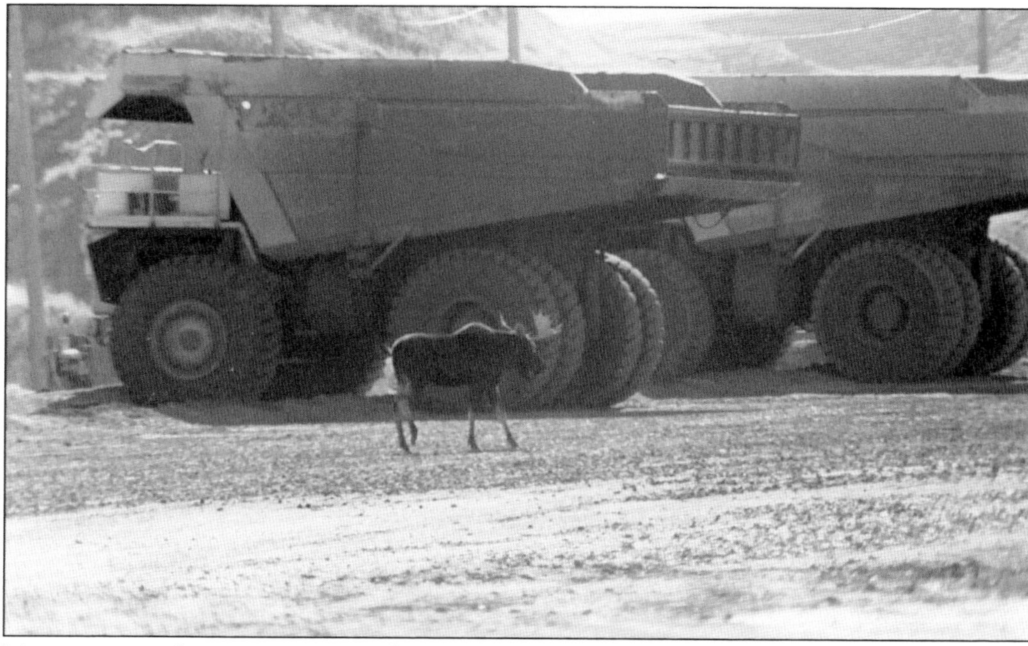

Moose are prevalent in many parts of Wyoming, but not usually on the prairie. The largest species of the deer family, they usually inhabit forests in subarctic climates. A male can stand 6.9 feet tall at the shoulder and weigh up to 1,500 pounds. Even as big as a moose is, this 170-ton haul truck is much bigger. (Courtesy of Buckskin Mine Employees.)

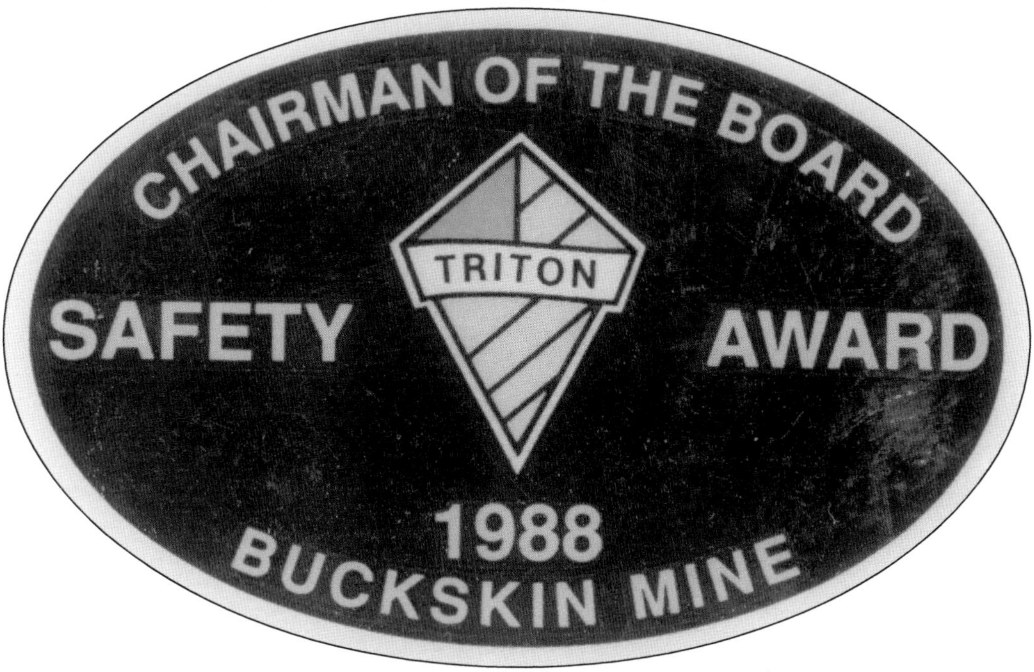

A subsidiary of Thunder Basin Coal Company, LLC, Triton Coal Company, LLC, operated two mines in the Powder River Basin: North Rochelle and Buckskin. In 2004, Arch Coal bought Triton's North Rochelle Mine, and Kiewit bought Buckskin Mine.

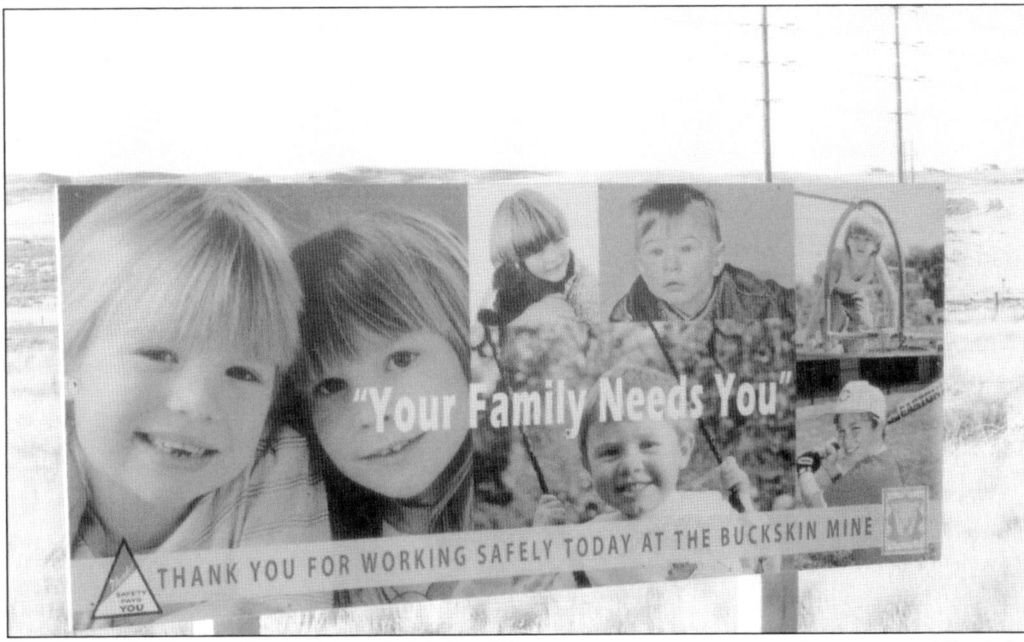

Buckskin Mine is recognized as one of the safest mining operations in the nation. All of the Campbell County mines employ dedicated safety professionals, and all employees are trained in proper safety. All new employees attend 40 hours of safety training prior to their first day on the job. Existing employees participate in annual safety refresher training. Every shift starts with walk-around safety inspections.

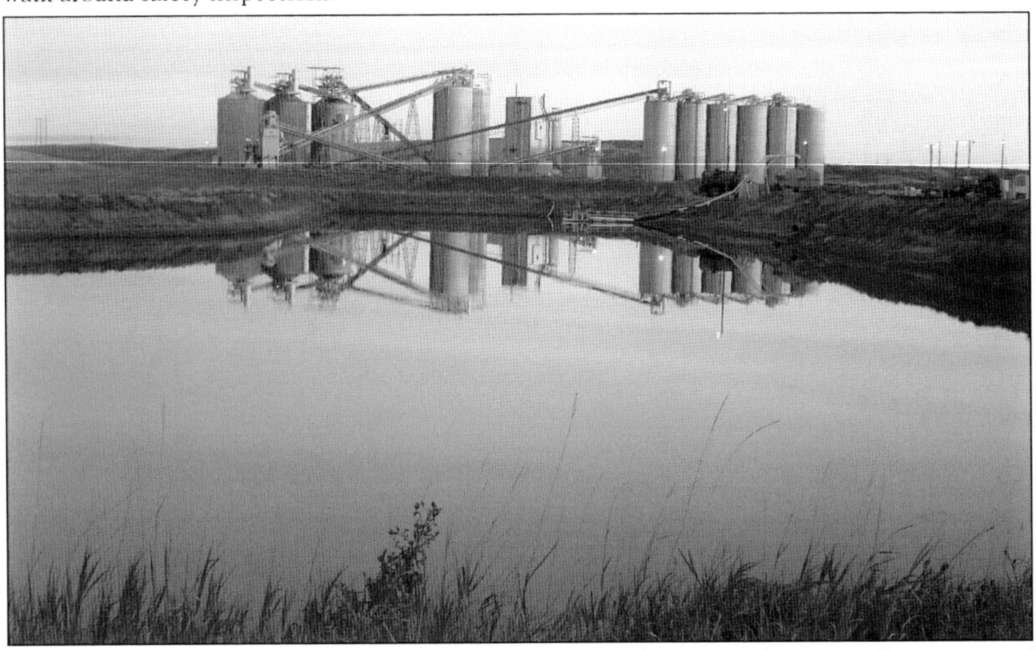

To the residents of Campbell County, a coal mine is beautiful. It provides an excellent wage and benefits for its workers, provides tax revenue, both locally and statewide, and keeps the lights on for millions of people throughout the nation. This photograph shows the beauty of Buckskin Mine, reflected in a storage pond.

Nine

Serving the Industry

For many years, Gillette was a small community that served the needs of the ranchers in the county. Businesses came and went, offering goods and services needed by the residents. Big Horn Tire began in Gillette in 1964 as East End Skelly, providing automobile tires. As they grew, they moved downtown and then to their current, larger location around 1972. It is no coincidence that that is the same time Belle Ayr Mine was under construction. Commercial tires for a variety of large equipment were added to the inventory at Big Horn Tires when the mines began production. In anticipation of employment, job seekers came to Campbell County by the thousands. Their first applications were at the coal mines, where starting pay for an apprentice heavy-equipment operator was more than $11 an hour, about what a cowhand would earn for half a day's work on a ranch. Gillette, once a cattle-loading stop on the Chicago, Burlington & Quincy Railroad, was quickly becoming an energy boomtown. Not only were private businesses growing, but Gillette also expanded its electrical power, sewage treatment, water supply and roads. New stick-built homes were out-numbering mobile homes. Campbell County's valuation grew along with everything else. The largest increases were in three areas—oil, coal production, and local property. Local property included new homes, commercial ventures and improvements at the coal mines. Also contributing to the county's rise in valuation were increases in the City, such as subdivision developments. The development of Wright, Wyoming, as housing for coal miners provided a large jump in valuation. Today, as shovels dig into the vast reserves of coal, companies are researching and developing new uses for the coal, and new ways to produce the coal that have less of an impact on the environment. The service category used to be welding and diesel fuel; the future category includes gasification, carbon capture and sequestration, and coal-to-liquids conversion.

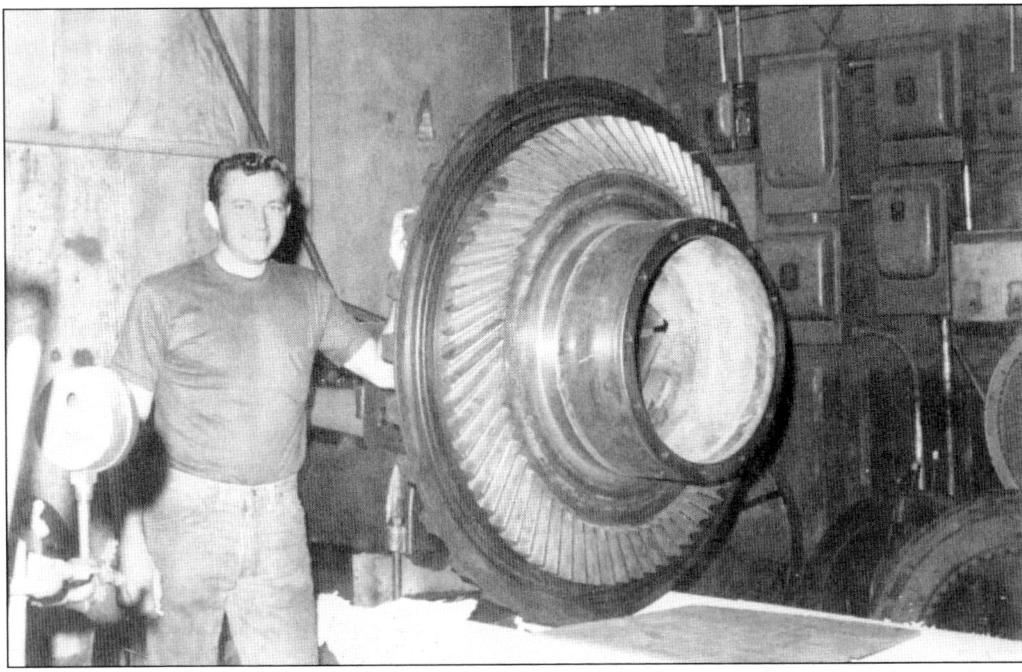

Draglines and other mining equipment are some of the world's largest and most complex machines. L&H Industrial is the world's leading supplier of aftermarket parts for draglines and shovels. The *L* in L&H stands for Leon Wandler, shown here working on a rotary drill table in the late 1960s. The *H* stands for Hank Barney, his partner and accountant when the business started in 1964. (Courtesy of L&H Industrial, Inc.)

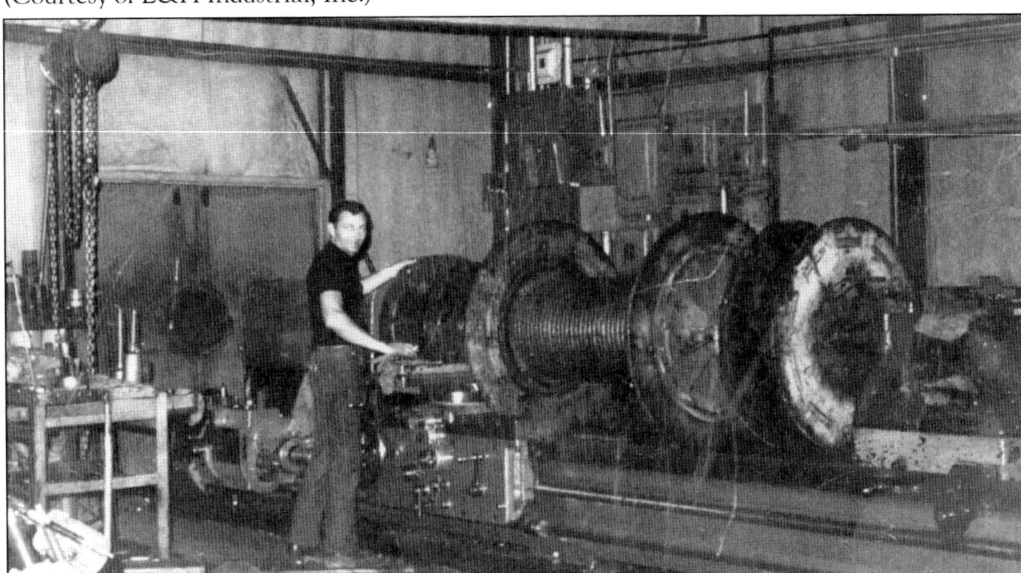

Leon Wandler and Hank Barney began as L&H Welding & Machine Company on Butler Spaeth Road. Their first focus was on oil field equipment, such as drilling rigs. When the oil boom busted in the 1980s, the focus went to repairing coal mine equipment. In the 1990s, Leon's sons joined their father. Leon is shown here, working on a Graw Works Drum in his shop in the late 1960s. (Courtesy of L&H Industrial, Inc.)

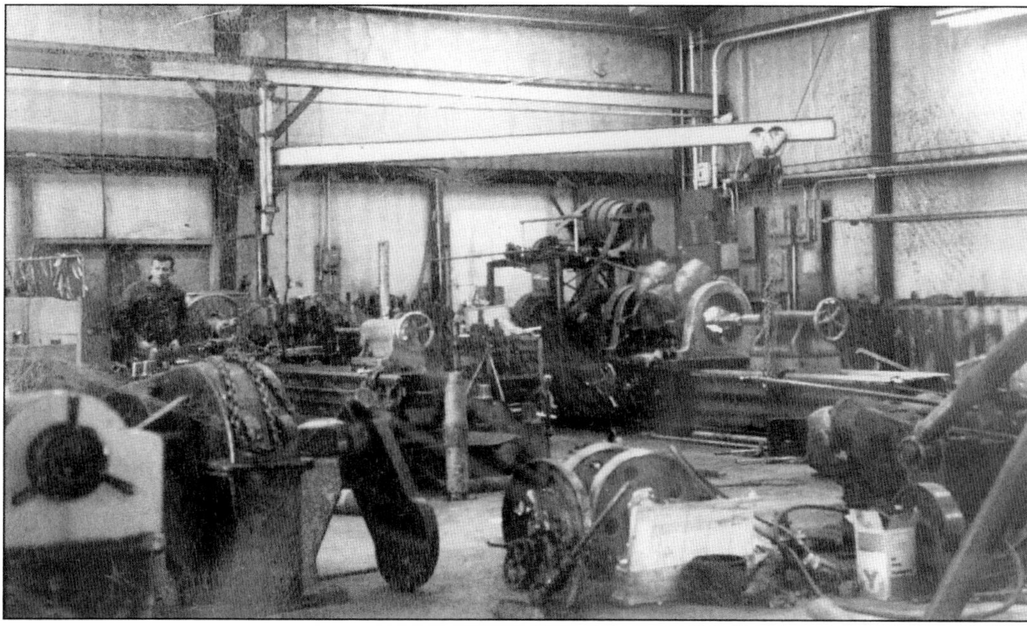

Core values of the business are honesty, integrity, loyalty, and respect. The company focuses on building long-term relationships with its customers. A recent bid to the National Aeronautics and Space Administration (NASA) has rewarded L&H with a contract for upgrades to crawler transporters at Kennedy Space Center. Shown here is Rick Wandler, working on a 12-inch Hollow Bore Spindle lathe cutting a drill collar connection end around 1968. (Courtesy of L&H Industrial, Inc.)

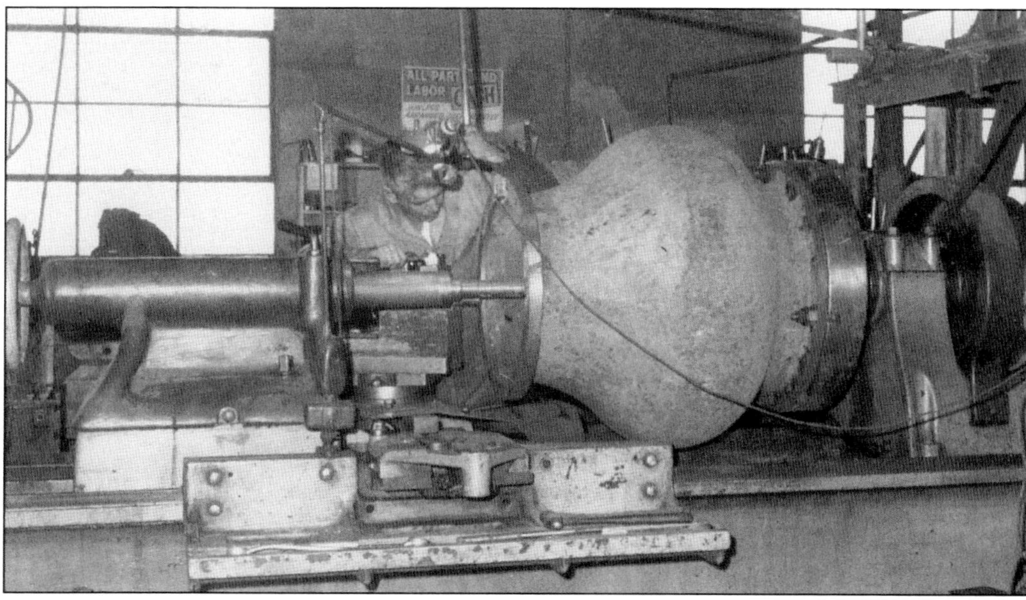

Caterpillar Inc., or "CAT," traces its origins to the 1925 merger of the Holt Manufacturing Company and the C.L. Best Tractor Company, creating the Caterpillar Tractor Company. In 1904, company photographer Charles Clements was watching an experimental tractor plow the land and said it crawled like a caterpillar. Mr. Holt jumped on the idea and said, "Caterpillar it is!" Caterpillar manufactures construction and mining equipment.

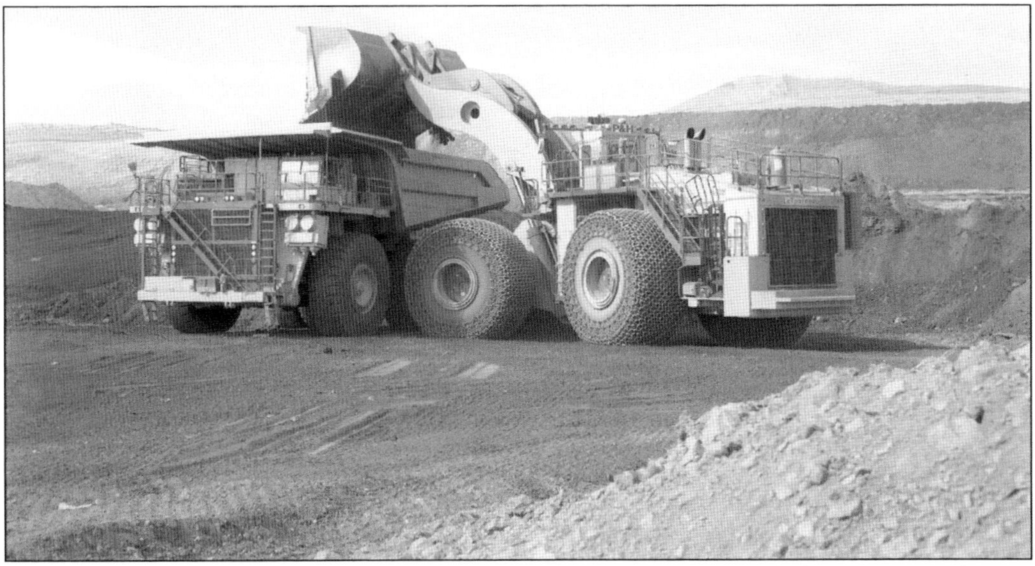

P&H Mining Equipment and LeTourneau, Inc. are subsidiaries of Joy Global, a mining machinery and service company. Alonzo Pawling and Henry Harnischfeger started the company in 1884 that would evolve into P&H Mining Equipment. Seventy percent of earthmoving equipment used during World War II was made by R.G. LeTourneau, Inc., another name synonymous with earthmoving equipment. This LeTourneau L2350 is the biggest front-end loader in the world. (Courtesy of Joy Global.)

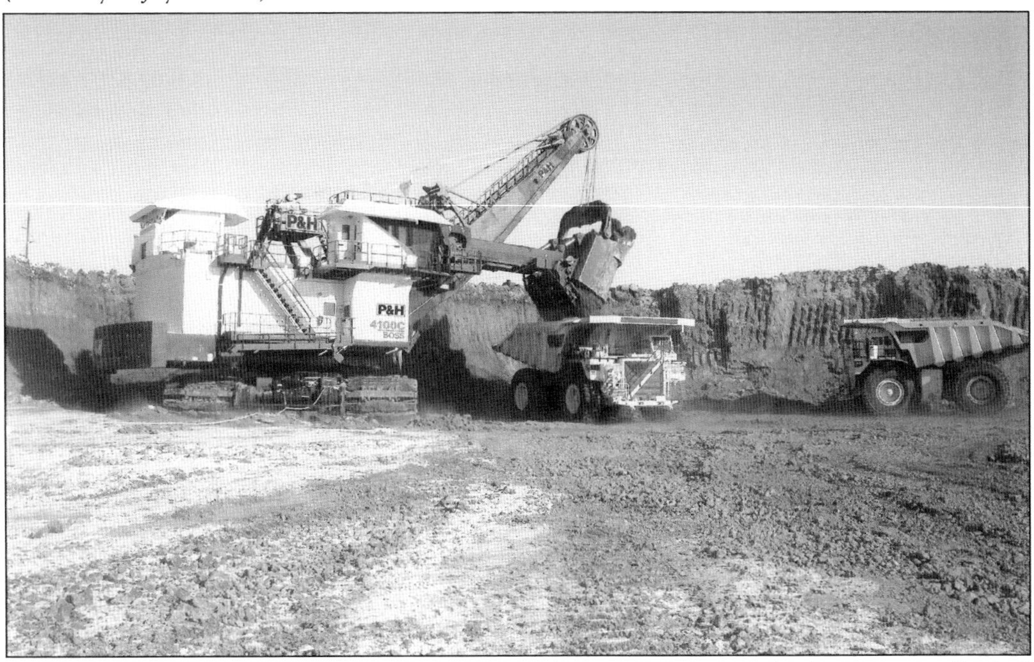

Shown here is a P&H 4100C Boss electric shovel with a 100-ton capacity loading a Caterpillar 793 haul truck, also with an electric drive. Mechanical drive has traditionally been the only option available, but when customers asked for equipment that was more reliable, easier to maintain, and friendlier to the operator and to the environment, equipment manufacturers responded with electric drives. (Courtesy of Joy Global.)

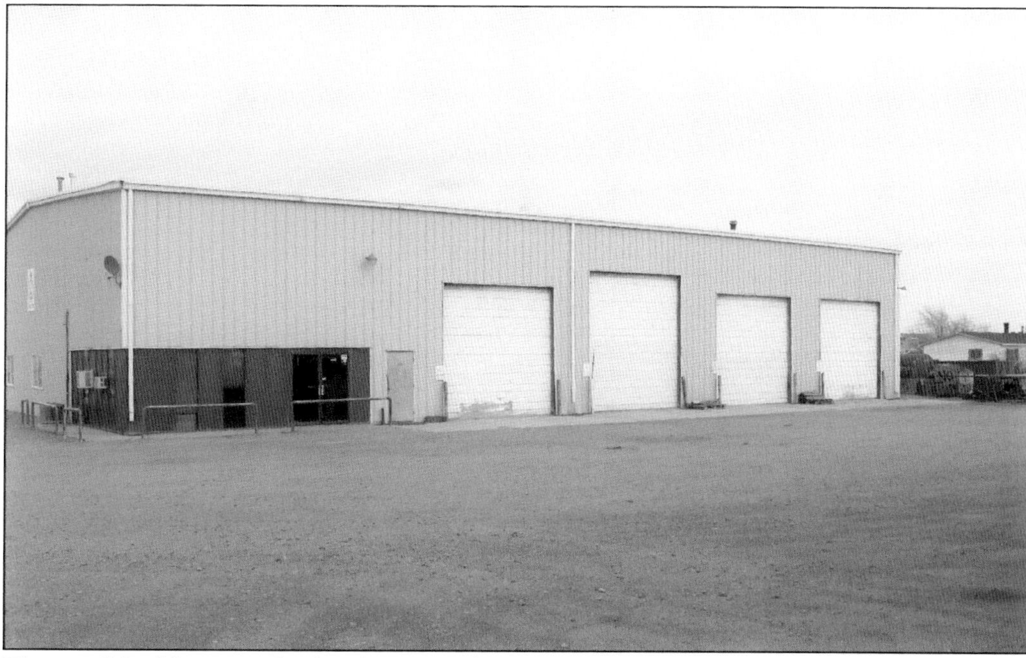

This small building was the first location of Holeman General Motors Diesel on Highway 14-16 in 1978. Holeman GM Diesel was purchased by Interstate Power Systems in 1984. Interstate was founded in 1957 as one of the world's first Detroit Diesel Engine distributors. (Courtesy of Interstate Companies, Inc.)

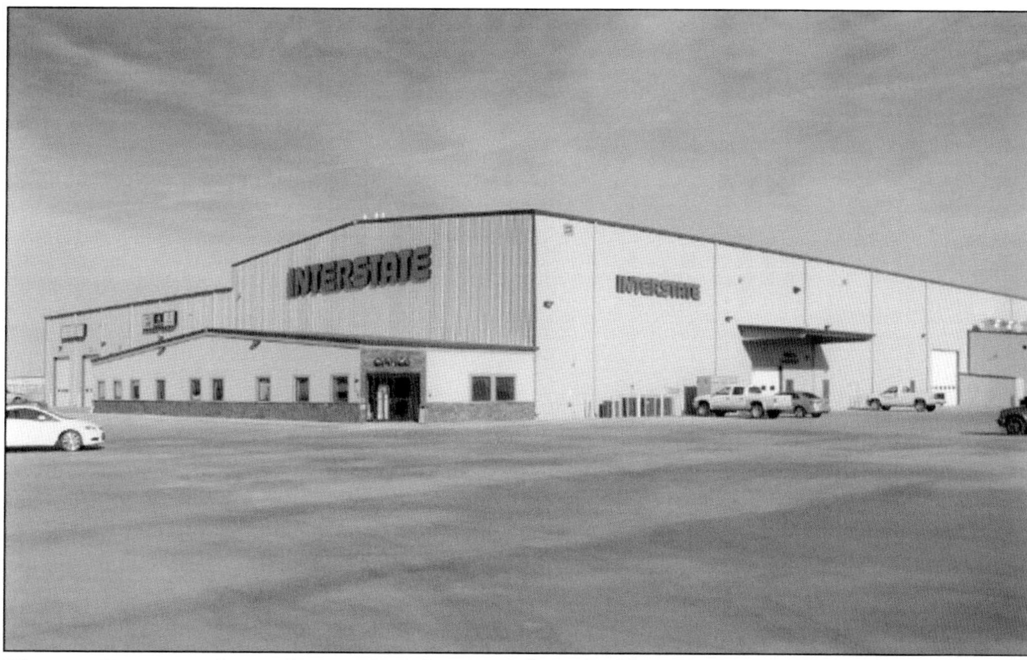

The new Interstate Power Systems building in Gillette held its grand opening and ribbon cutting during 2012. Besides providing power generation and engines, Interstate services and repairs conveyor systems in place at most mines in the Powder River Basin. (Courtesy of Interstate Companies, Inc.)

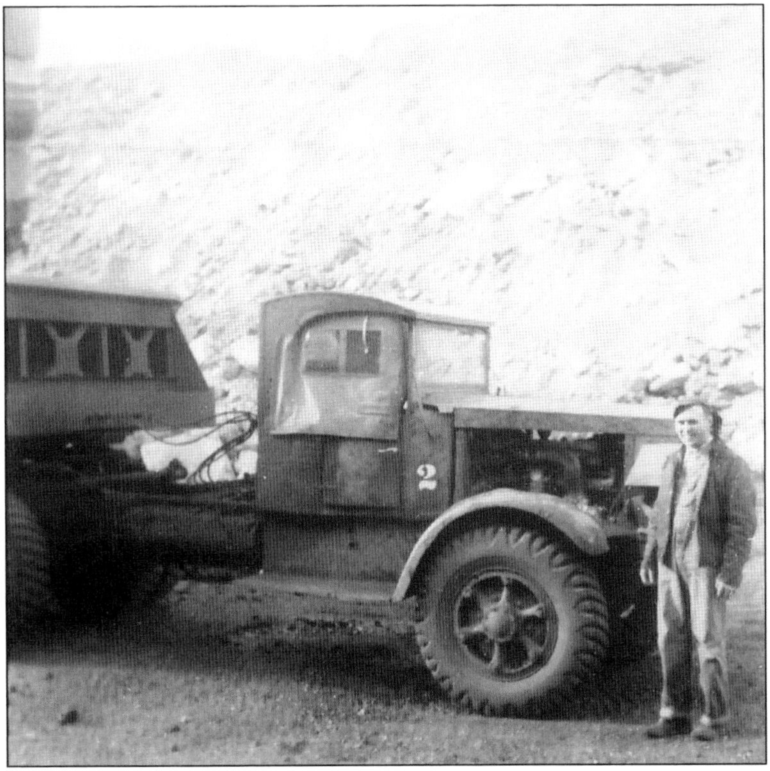

Clessie Cummins was a visionary mechanic. William Glanton Irwin was a businessman. In 1919, they partnered to form Cummins Engine Company, now Cummins, Inc. Clessie Cummins was the first to see the commercial potential of an unproven engine technology invented two decades earlier by Rudolph Diesel. After a Christmas tour in the first diesel-powered automobile, Cummins was able to gain a foothold as an engine supplier to the trucking industry. (Courtesy of Cummins Inc.)

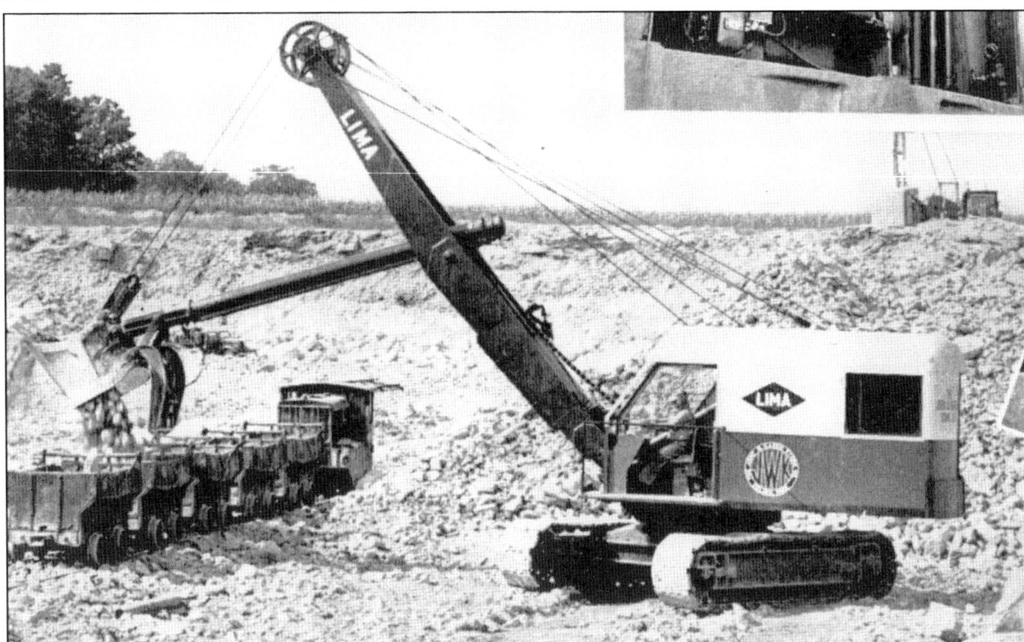

Cummins had a fondness for races at the Indianapolis Motor Speedway. His diesel engines were field tested in many Indy cars. From these humble beginnings, Cummins grew to open its first foreign manufacturing facility in Scotland in 1956 and today has more than 5,000 facilities in 197 countries, and provides jobs for 40,000 employees. (Courtesy of Cummins Inc.)

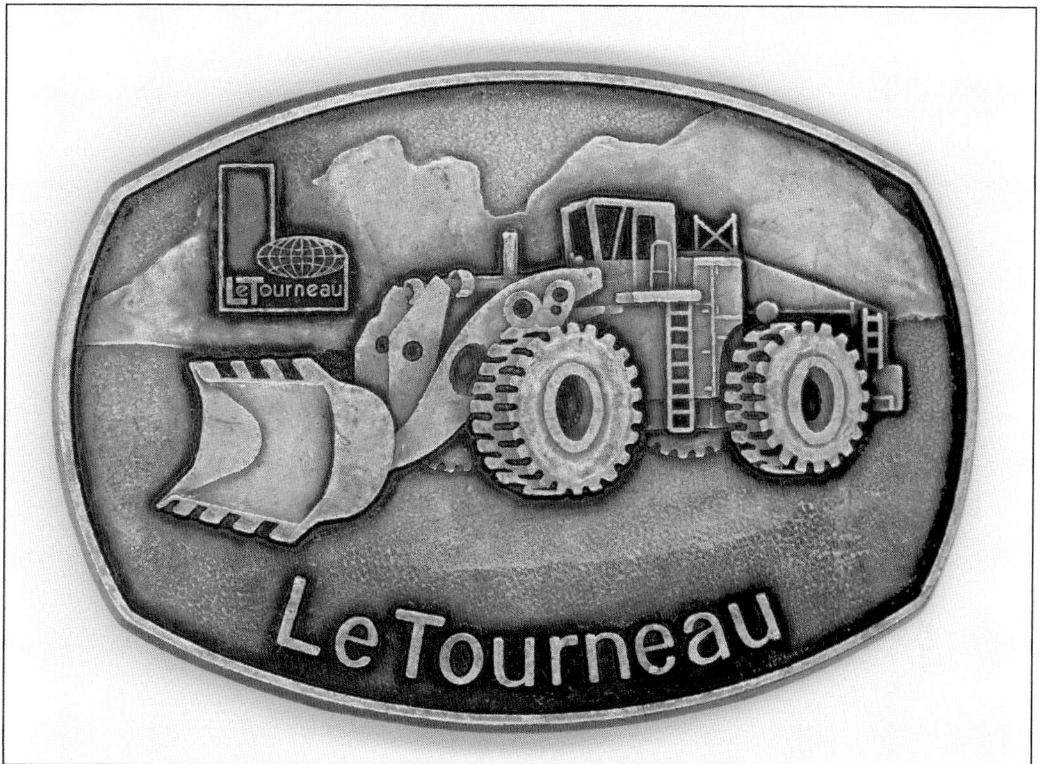

LeTourneau is a world leader in manufacturing large-capacity wheel loaders, energy-efficient, high-performance, diesel electric loaders, and dozers for the surface mining industry, as well as equipment for the oil and gas drilling industries. Joy Global recently acquired LeTourneau. (Courtesy of Joy Global.)

Wyoming Machinery originally started in Casper in 1969, expanding to Cheyenne in 1972, and Gillette in 1978, in response to customer needs. Wyoming Machinery supports three industries: mining, construction, and oil and gas. As the only Caterpillar dealer in Wyoming, they sell and service the world's largest manufacturer of construction and mining equipment that dates back to the early 1900s.

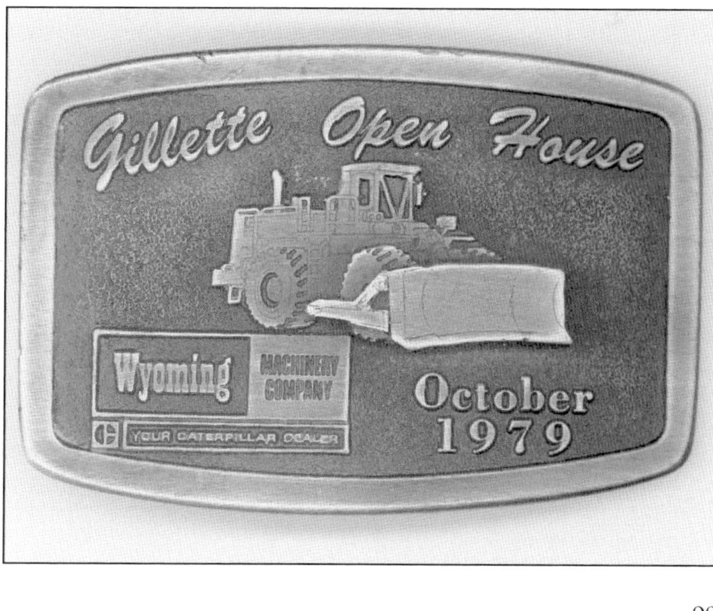

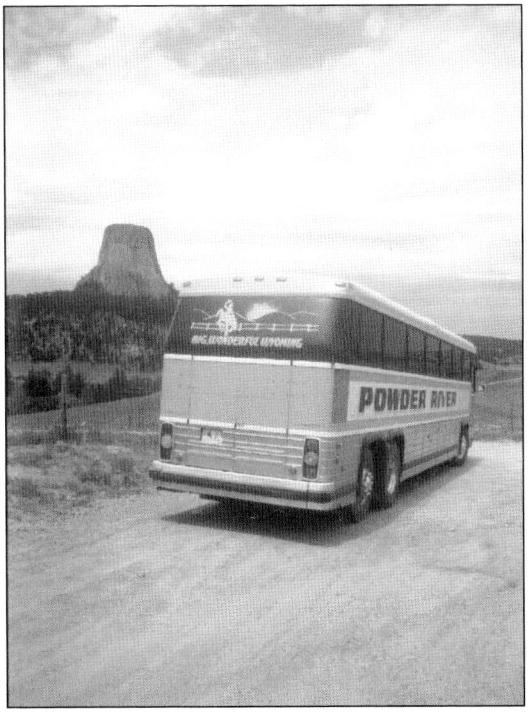

Worthen Van Service was started by Gillette resident Wayne Worthen in the late 1970s, to offer transportation services to railroad workers and coal miners who worked up to 65 miles south of Gillette. The business grew to offer charter bus service and, for many years, public transportation throughout Wyoming. Today, Powder River Transportation is a Coach America company. (Courtesy of Powder River Transportation.)

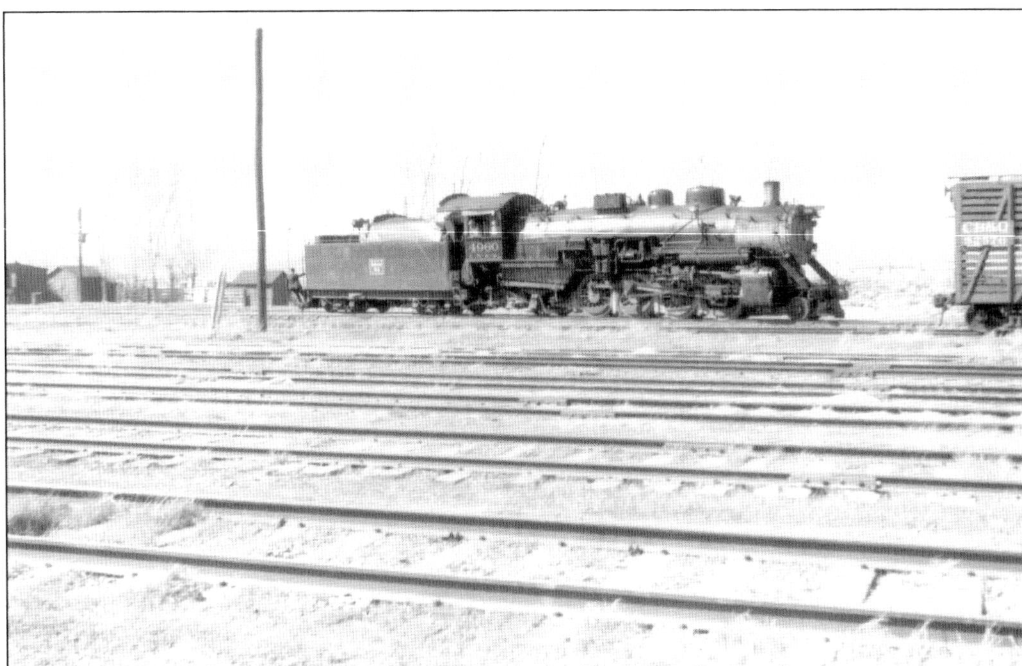

Railroad tracks were laid into Gillette in August 1891, forever changing the way stockmen did business in northeastern Wyoming. Engineer and surveyor Edward Gillette was in charge of surveying the area for the Burlington & Missouri Railroad. Gillette saved the railroad five miles of track and 30 bridges, for which the railroad named the new town after him. This photograph is of the last steam locomotive used in Gillette.

The last passenger train through Gillette is parked in front of the Wyodak power plant in 1969. Passenger trains stopped being a cost-effective means of transportation; however, coal trains soon filled the railroad tracks. Rail spurs were run from the main line to mine properties as part of the mine's development.

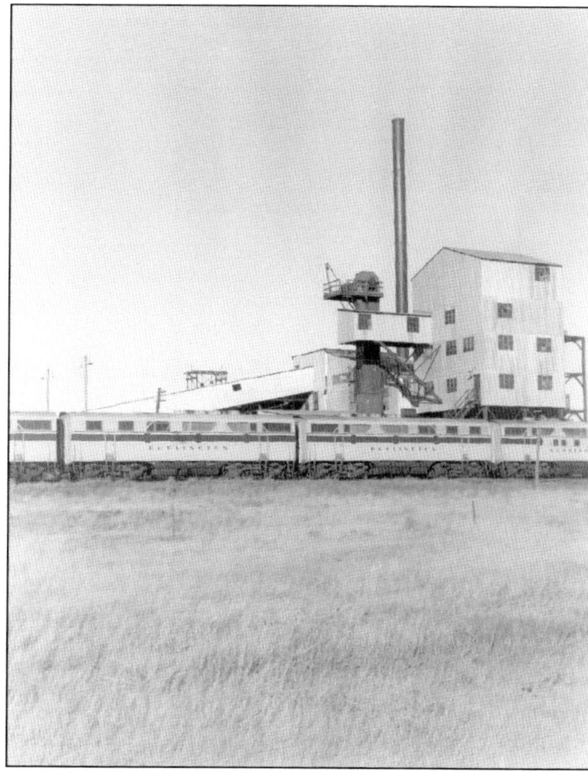

The final section of track is lowered to link Union Pacific System with Chicago & Northwestern's new 107-mile connector line to Wyoming's coal-rich Powder River Basin. Union Pacific workers made the final link on July 11, 1984, near Gering, Nebraska. This joint line runs from Caballo Mine to Shawnee, Nebraska, and is controlled by the Burlington Northern Santa Fe.

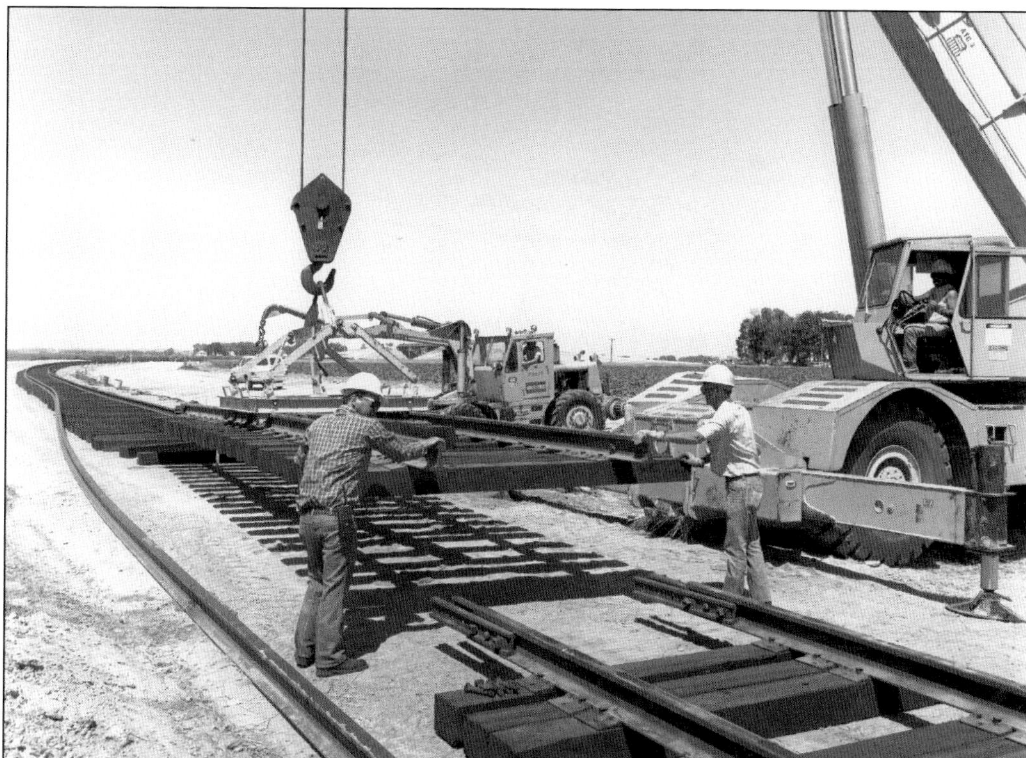

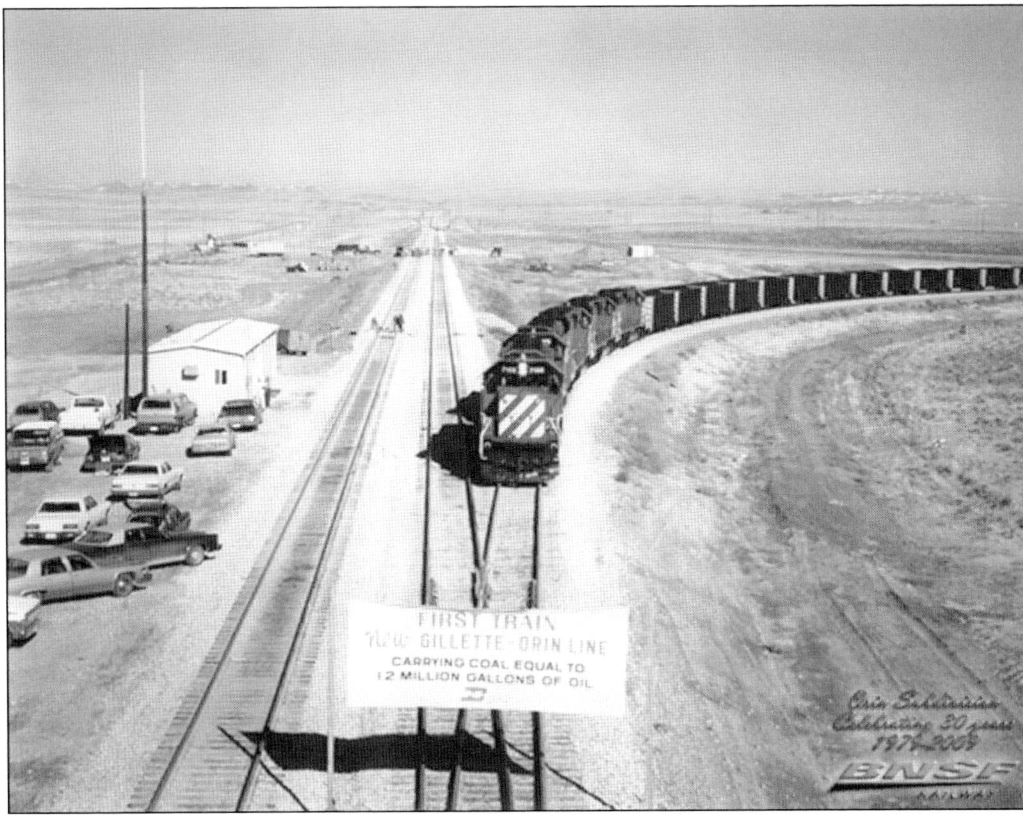

This photograph shows the first train on the new Gillette to Orin line carrying coal equal to 12 million gallons of oil. Early railcars, averaging 100 tons per car, were made of steel. Modern railcars are aluminum and carry approximately 120 tons per car. Train lengths average 135 to 150 cars, depending on the utility receiving the train. Not all utilities can handle a train that long. (Courtesy of Burlington Northern Santa Fe Railroad.)

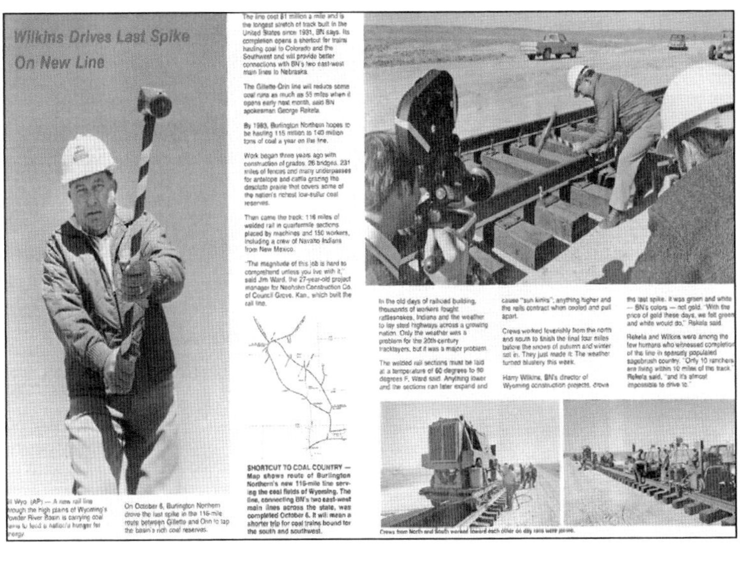

Wilkins Drives Last Spike On New Line

...Wyo. (AP) — A new rail line ...rough the high plains of Wyoming's ...owder River Basin is carrying coal ... to feed a nation's hunger for ...nergy.

On October 6, Burlington Northern drove the last spike in the 116-mile route between Gillette and Orin to tap the basin's rich coal reserves.

The line cost $1 million a mile and is the longest stretch of track built in the United States since 1931, BN says. Its completion opens a shortcut for trains hauling coal to Colorado and the Southwest and will provide better connections with BN's two east-west main lines to Nebraska.

The Gillette-Orin line will reduce some coal runs as much as 55 miles when it opens early next month, said BN spokesman George Rekela.

By 1983, Burlington Northern hopes to be hauling 115 million to 140 million tons of coal a year on the line.

Work began three years ago with construction of grades, 26 bridges, 231 miles of fences and many underpasses for antelope and cattle grazing the desolate prairie that covers some of the nation's richest low-sulfur coal reserves.

Then came the track: 116 miles of welded rail in quartermile sections placed by machines and 150 workers, including a crew of Navaho Indians from New Mexico.

"The magnitude of this job is hard to comprehend unless you live with it," said Jim Ward, the 27-year-old project manager for Neosho Construction Co. of Council Grove, Kan., which built the rail line.

SHORTCUT TO COAL COUNTRY — Map shows route of Burlington Northern's new 116-mile line serving the coal fields of Wyoming. The line, connecting BN's two east-west main lines across the state, was completed October 6. It will mean a shorter trip for coal trains bound for the south and southwest.

In the old days of railroad building, thousands of workers fought rattlesnakes, Indians and the weather to lay steel highways across a growing nation. Only the weather was a problem for the 20th-century tracklayers, but it was a major problem.

The welded rail sections must be laid at a temperature of 60 degrees to 90 degrees F, Ward said. Anything lower and the sections can later expand and cause "sun kinks"; anything higher and the rails contract when cooled and pull apart.

Crews worked feverishly from the north and south to finish the final four miles before the snows of autumn and winter set in. They just made it. The weather turned blustery this week.

Harry Wilkins, BN's director of Wyoming construction projects, drove the last spike. It was green and white — BN's colors — not gold. "With the price of gold these days, we felt green and white would do," Rekela said.

Rekela and Wilkins were among the few humans who witnessed completion of the line in sparsely populated sagebrush country. "Only 10 ranchers are living within 10 miles of the track," Rekela said, "and it's almost impossible to drive to."

Crews from North and South worked toward each other on day rails were joined.

The Union Pacific cannot go north past the Caballo junction, meaning the mines north of Gillette—Eagle Butte, Buckskin, and Rawhide—are served only by the Burlington Northern. The southern mines can be served by either railroad. The tracks that are served by both railroads are called "the joint line." (Courtesy of Burlington Northern Santa Fe Railroad.)

The colors of the Burlington Northern trace its history to the Chicago, Burlington, and Quincy Railroad. Locomotives were painted green, black, and white when Burlington Northern introduced its logo. Unfortunately, the locomotives and cars had an unfortunate habit of camouflaging into the scenery and causing accidents at railroad crossings. After several visibility studies, additional white was added to the front, which was very successful. (Courtesy of Burlington Northern Santa Fe Railroad.)

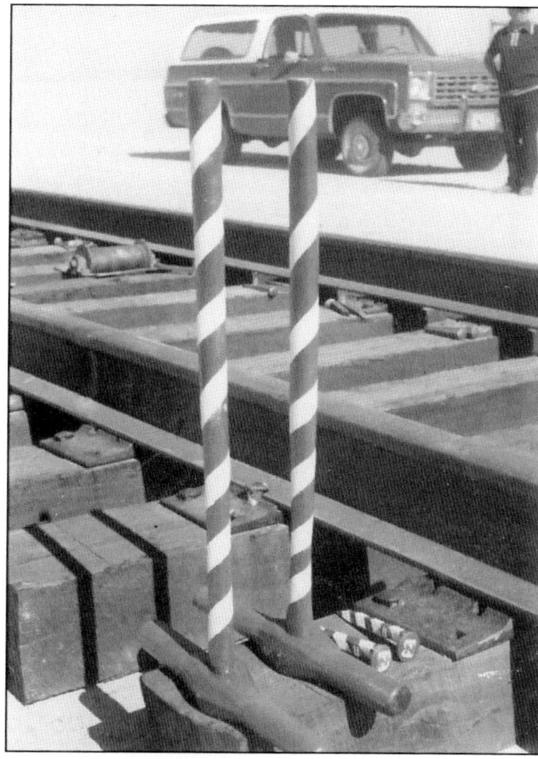

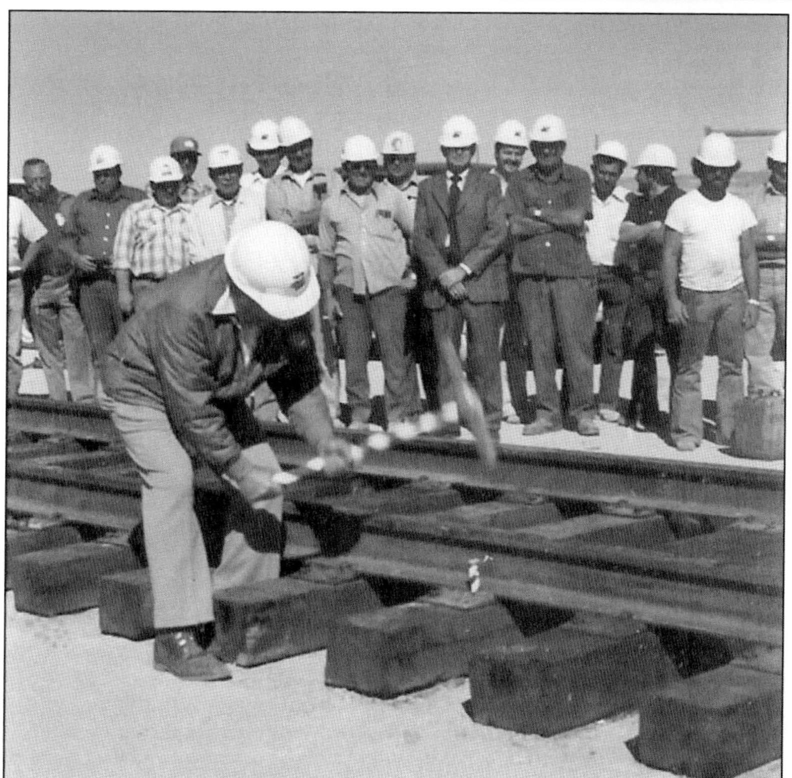

On October 6, 1979, the last spike was driven into the tracks of the new 116-mile route between Gillette and Orin. The spike was not gold but Burlington Northern green. The line cost $1 million per mile. This line is the longest track built in the United States since 1931. The Orin Line cost $110 million and provides a shortcut to the main lines to Colorado, Texas, and Nebraska. (Courtesy of the Burlington Northern Santa Fe Railroad.)

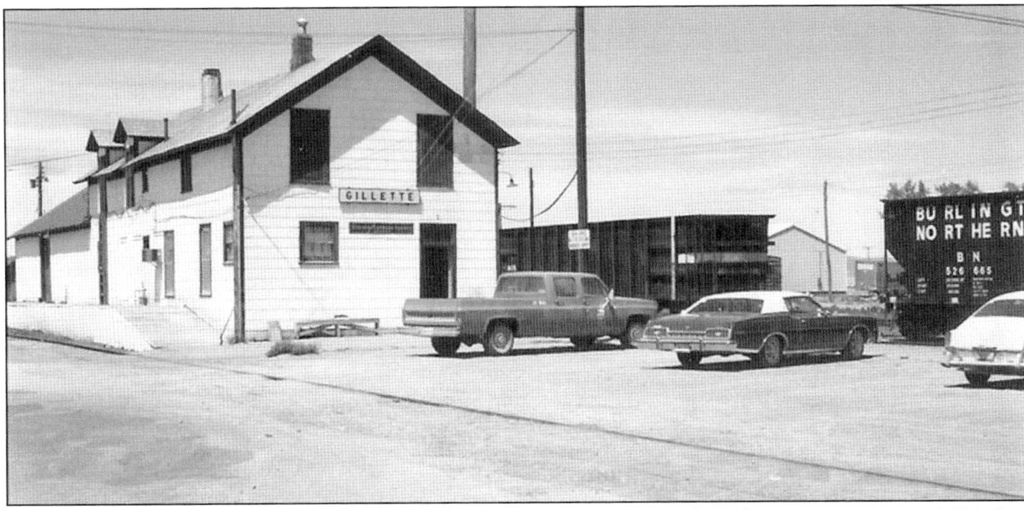

The old depot at Gillette had most likely been there since the railroad came to town in 1891. It is shown in many historical photographs with a fresh coat of white paint from time to time. There were also no telephones in the early days; all communication was done by telegram, which was handled by the Burlington railroad at the depot. The new yard office was built in 1979. (Courtesy of Burlington Northern Santa Fe Railroad.)

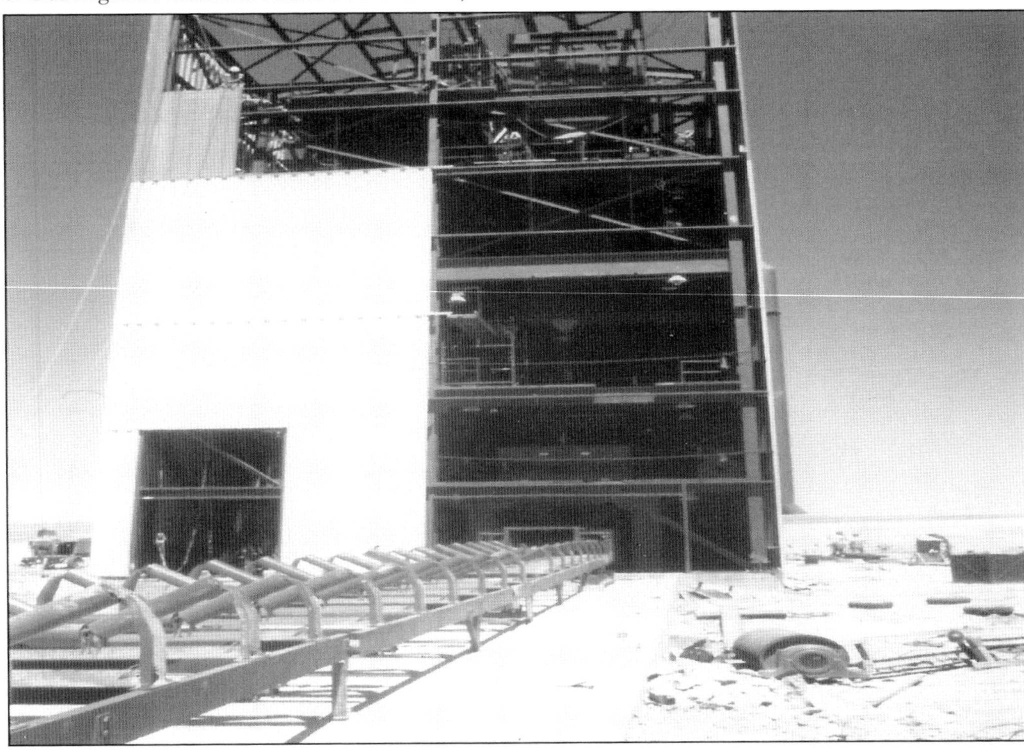

Coal trains make up the majority of the rail traffic in Campbell County, but several freight trains also pass through each day. Coal from the Powder River Basin is mined at the rate of 12 tons per second and is shipped to 37 states. Loadout facilities, like this one, were built in response to demands for more accurate weighing. Loadout facilities can get within 250 pounds of consistent weight in railcars. (Courtesy of Burlington Northern Santa Fe Railroad.)

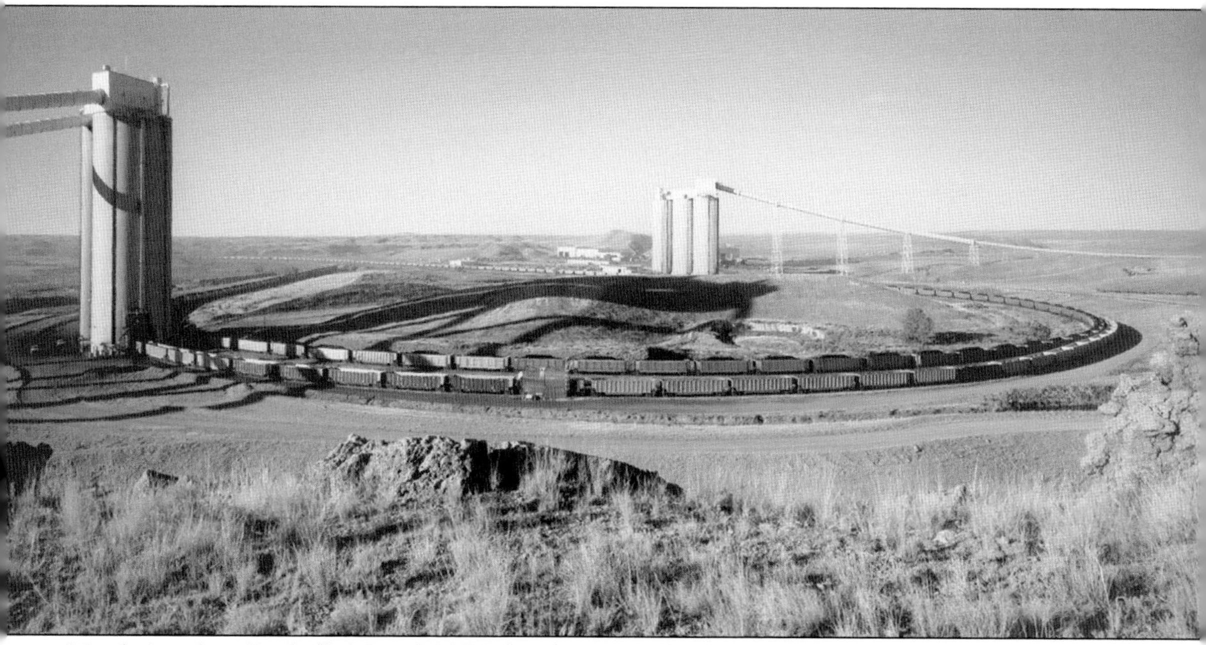

North Antelope Rochelle Mine (NARM) is the setting for this photograph. Two concentric loop tracks connect with the Burlington Northern Santa Fe and Union Pacific railroads' joint line. The loop tracks are capable of handling 150 cars per unit train. A unit train is made up of railcars carrying a single type of commodity, all bound for the same destination. By hauling only one kind of freight for one destination, a unit train does not need to switch cars at various junctions, and so can make nonstop runs between two terminals. This reduces not only the shipping time but also the cost. The unit train was introduced by railroad companies in the 1950s so they could offer lower shipping rates and make their freight service more marketable. Initially, unit trains were used primarily to haul coal from the mines to power plants. By the late 20th century, about 50 percent of the coal shipped in the United States was shipped by unit trains. Other forms of bulk cargo, such as grain and cement, were also transported in this way. (Courtesy of Burlington Northern Santa Fe Railroad.)

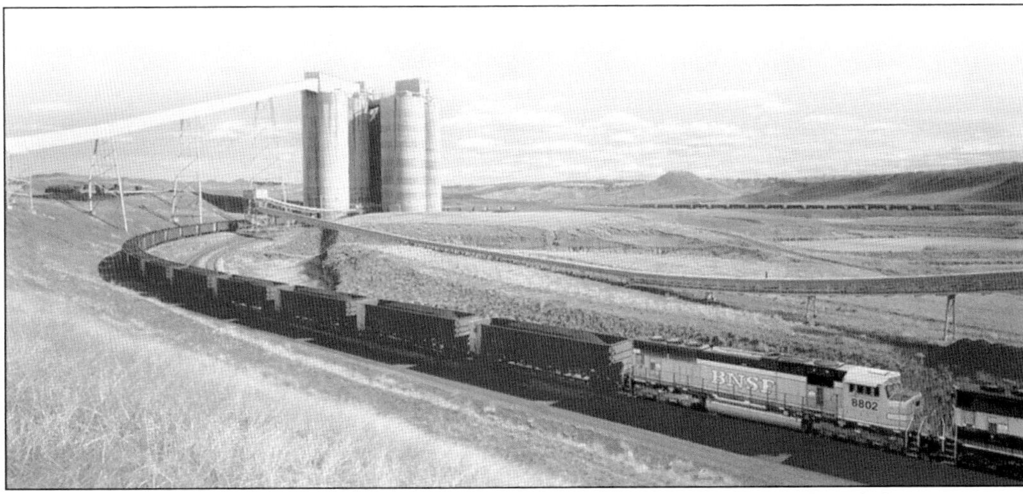

At Rawhide Mine, coal is hauled to an in-pit truck dump, where it is sized and then conveyed to a 12,000-ton intermediate silo. From there, coal is conveyed to secondary crushers for final sizing. Then it is placed in loadout silos that have a combined capacity of 75,000 tons. (Courtesy of Burlington Northern Santa Fe Railroad.)

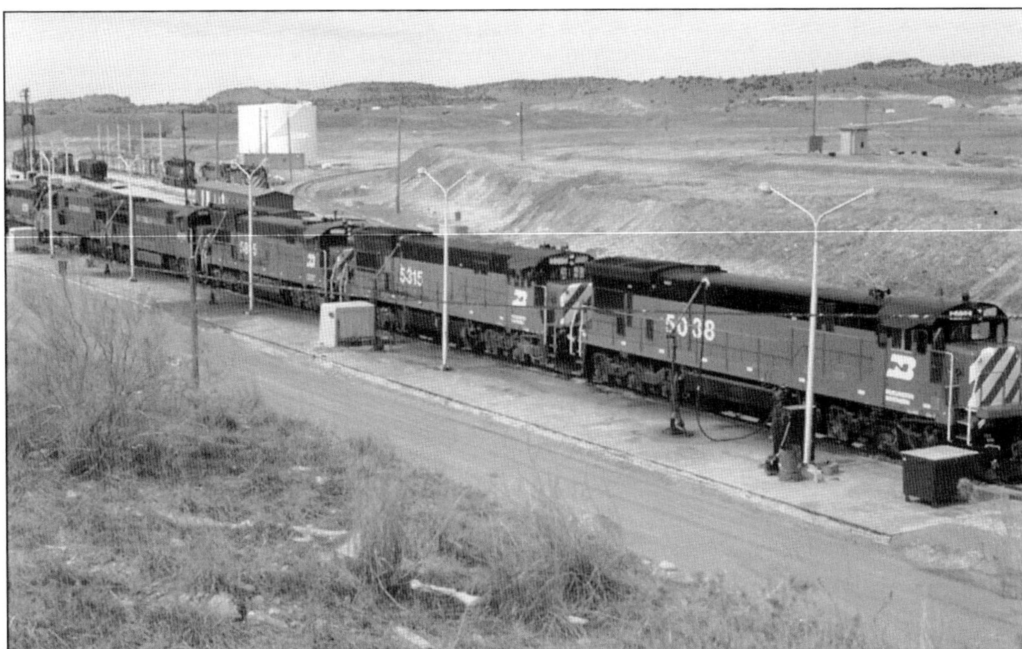

A Burlington Northern train stops at a fueling station along the route through the Wyoming prairie. Throughout its history, the Burlington Northern had various interesting test paint schemes. One was the High Visibility tiger stripes; one was to honor the employees serving in Desert Storm; and one commemorated the purchase of the St. Louis–San Francisco line in 1980. (Courtesy of Burlington Northern Santa Fe Railroad.)

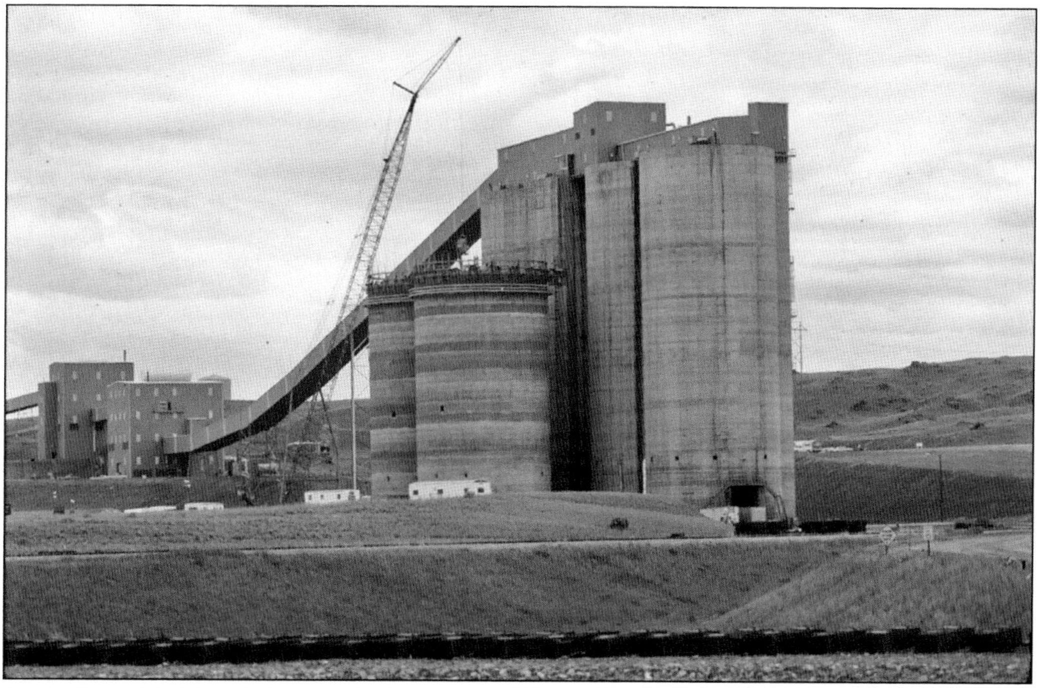

At the Rawhide Mine dedication in September 1977, Ed Smith Jr., president of The Carter Mining Company, said when Rawhide Mine achieved its production goal of 12 million tons of coal a year, it would be producing enough energy to supply electricity for over 1.7 million homes. That amount would be equal to 100,000 barrels a day of oil production. Rawhide has produced 12 million tons a year many times. (Courtesy of Burlington Northern Santa Fe Railroad.)

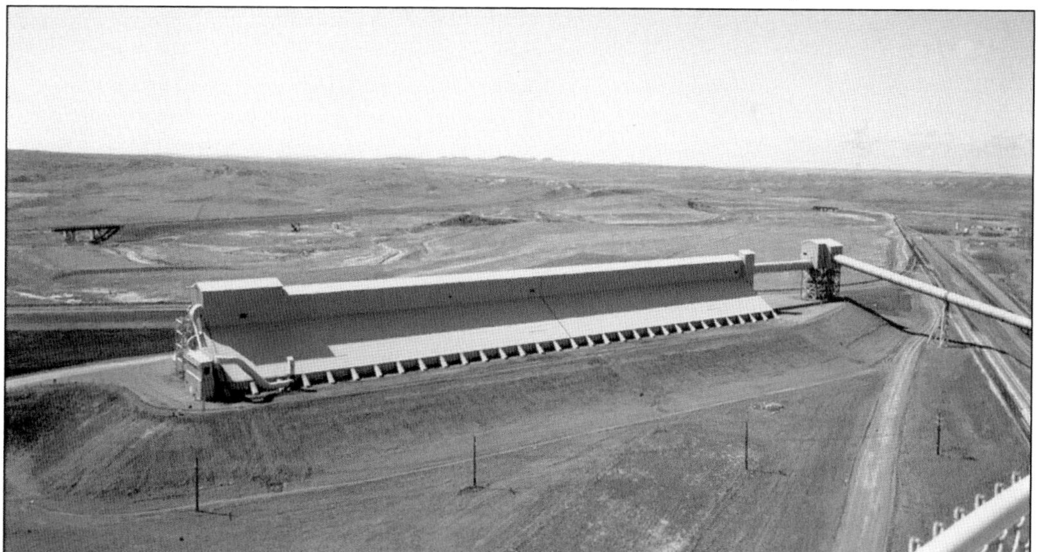

This long building at Cordero Mine is called a slot storage bin. It was built with a 100,000-ton capacity to help the mine continue to produce coal if the four silos were full. It can also be used for blending to meet a customer's specifications. When coal in the slot storage bin is needed, an "arm" inside pushes the coal onto a conveyor, where it travels to the loadout silos. (Courtesy of Burlington Northern Santa Fe Railroad.)

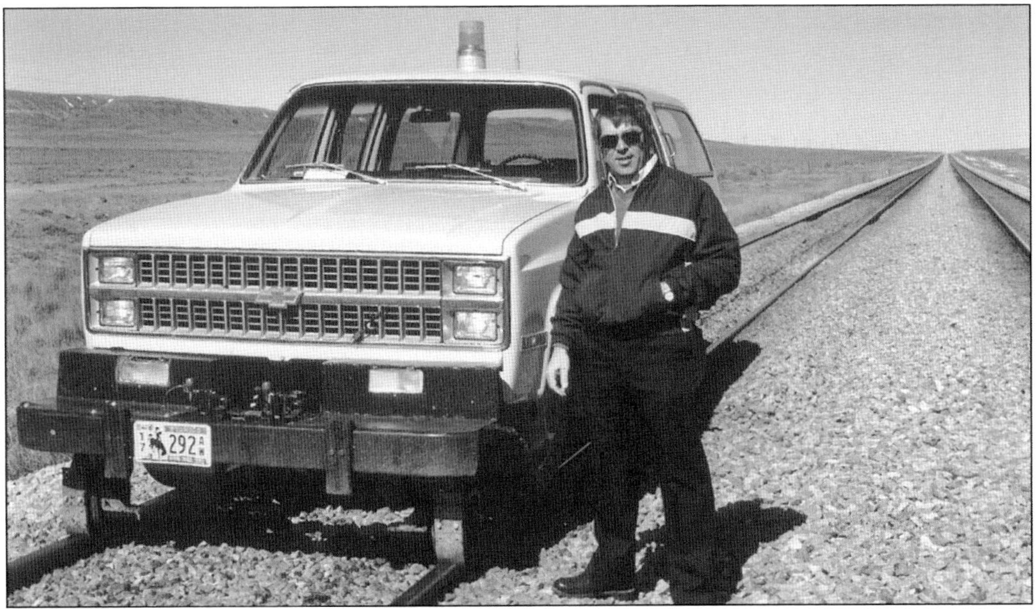

A road-rail vehicle can operate both on rail tracks and a traditional road. They are often road vehicles keeping their normal wheels with rubber tires but fitted with additional steel wheels for running on rails. The rail wheels are raised and lowered as needed. Shown here is Jake Greeling, former superintendent of operations. (Courtesy of Burlington Northern Santa Fe Railroad.)

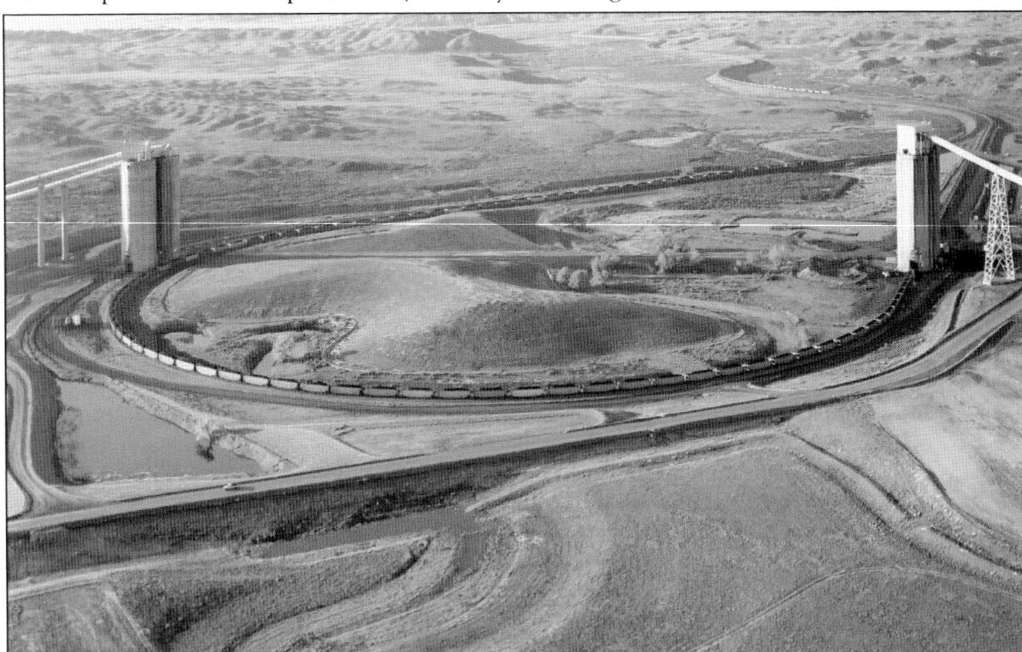

Another view of NARM shows its isolated location on the high plains. NARM—the world's largest and most productive coal mine—was recognized with a US Department of the Interior 2012 Excellence in Surface Coal Mining Reclamation Award for its continuing efforts to protect eagles, hawks, owls, and other birds of prey by creating habitat and protecting nesting areas. (Courtesy of Burlington Northern Santa Fe Railroad.)

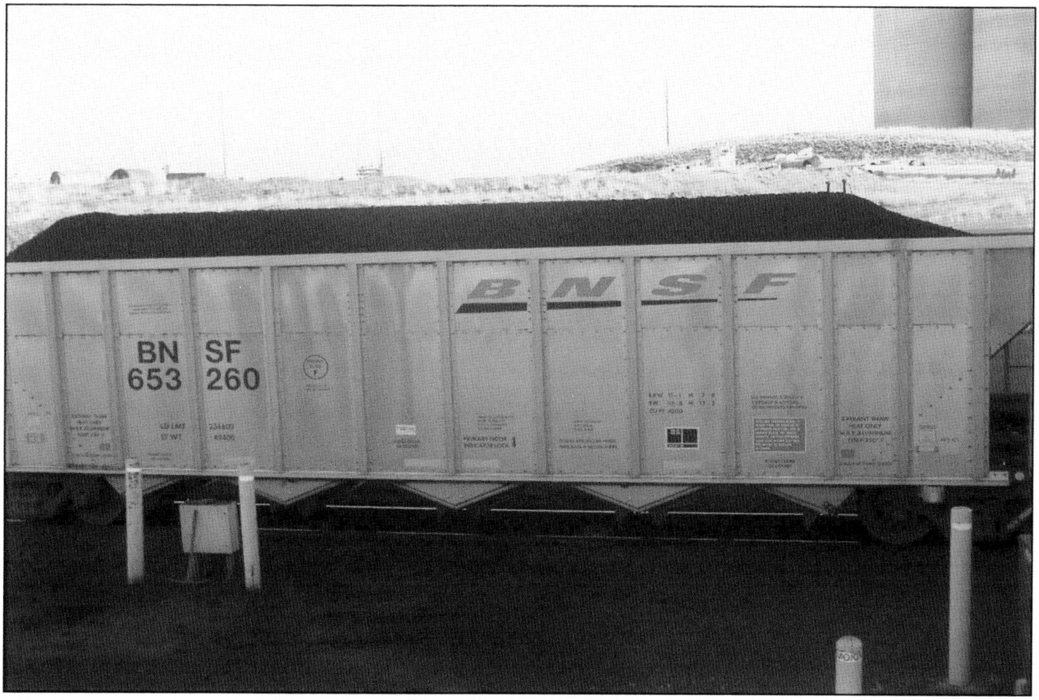

The letters on the side of a railcar reflect who owns the car and, sometimes, their destination. Some of the contracts filled with Powder River Basin coal are Georgia Power (letters RWSX) and Dairyland Power Cooperative (letters DAPX). The X at the end of the letters indicates that the utility owns the cars. This car is owned by the Burlington Northern Santa Fe Railroad, with no markings to indicate which utility gets this coal. (Courtesy of Burlington Northern Santa Fe Railroad.)

Railcars can be treated inside with a release agent to prevent frozen coal from remaining in the car after unloading; the coal itself can be sprayed with a solution to suppress dust. Orders for either of these originate with the utility buying the coal.

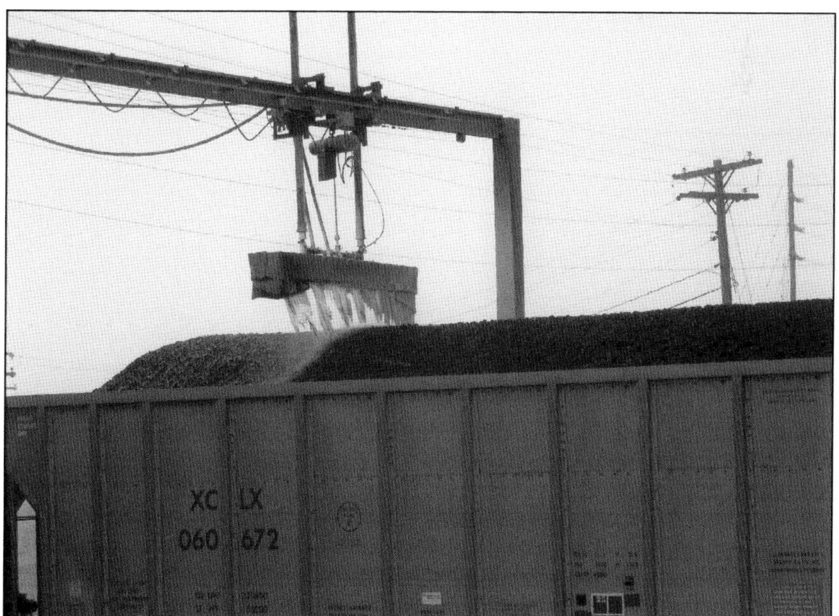

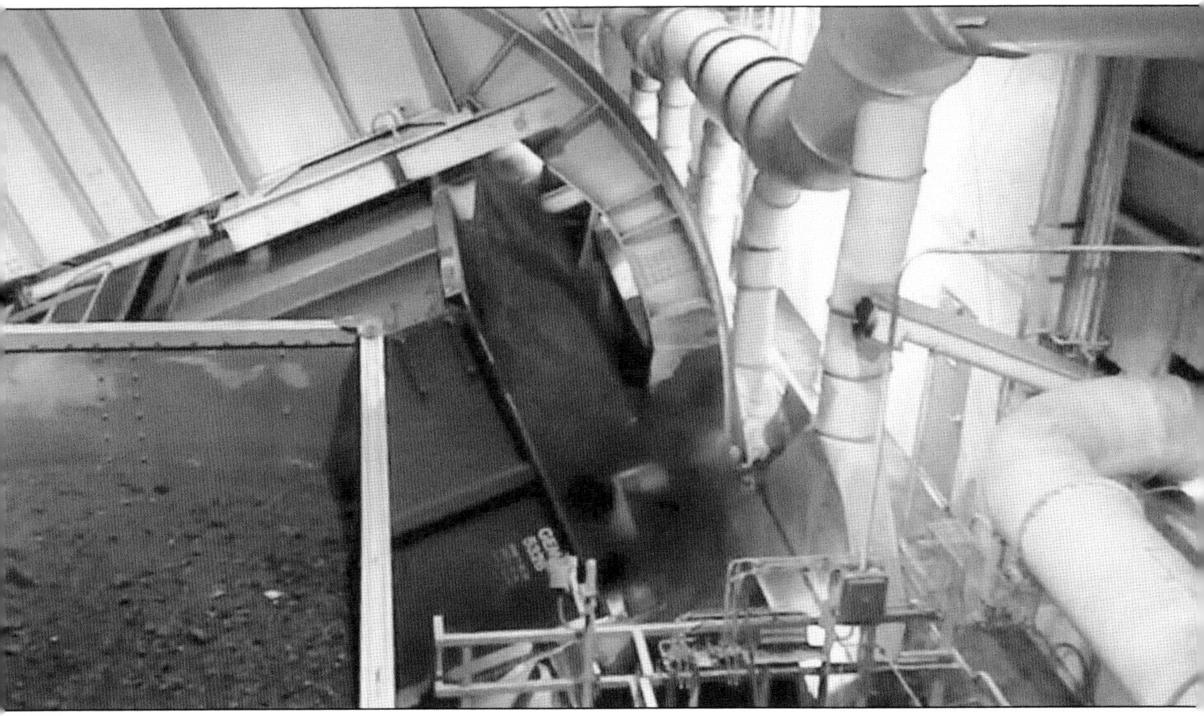

As railroad cars evolved, there needed to be a more efficient way to unload bulk materials from them than to use a shovel. A debate began about whether it was better to turn the whole car upside-down or allow the bulk materials to fall out of the bottom. The first type is called a rotary dump, such as the car shown here. The train enters the utility, where it is uncoupled, secured to a rotary dumper, turned over until empty, and put back in place on the train. It eventually occurred to someone that if the railcar was rotated about its own coupler, there would be no need to uncouple the cars to dump them. This led to the development of the rotating coupling. The second type opens underneath, known as the belly or the bottom, to let the coal fall out. The destination determines if they receive rotary dump or bottom dump cars.

Ten

Safety Is No Accident

Safety is of the utmost importance at all Powder River Basin mines and service providers. Sometimes, things do not end up the way they were planned, but not for a lack of processes in place. Improved equipment and increased training have contributed to enhanced mine operations over the years. The goal is always no LTAs—lost time accidents. One procedure that helps mine safety is requiring vehicles on mine property to drive on the left-hand side of the road; this procedure especially helps the huge haul trucks. There are two reasons for this: It gives the haul truck driver a better view of the edge of the road from his or her vantage point in the cab which is mounted on the left side of the truck, and it keeps drivers in two haul trucks as far apart from each other as possible should they "touch." In 1891, Congress passed the first federal statute governing mine safety, which provided for minimum ventilation requirements at underground mines and prohibited operators from employing children less than 12 years of age. The United States Bureau of Mines was created on May 16, 1910, to deal with a wave of catastrophic mine disasters. Its mission was to conduct research to enhance the safety, better the health, and decrease the environmental impacts of mining and processing of minerals, and to share information about mine safety around the world. After a series of additional safety acts, the Federal Coal Mine Health and Safety Act of 1969 was passed. This act required two annual inspections at surface mines and provided for monetary and criminal penalties for violations. In 1973, the secretary of the interior created the Mining Enforcement and Safety Administration (MESA) as a new agency separate from the Bureau. Next, Congress passed the Federal Mine Safety and Health Act of 1977 (Mine Act) which strengthened and expanded the rights of miners and changed the name of MESA to the Mine Safety and Health Administration (MSHA). With additional emphasis on safety, yearly mining fatalities nationwide dropped sharply from 272 in 1977, to 86 in 2000.

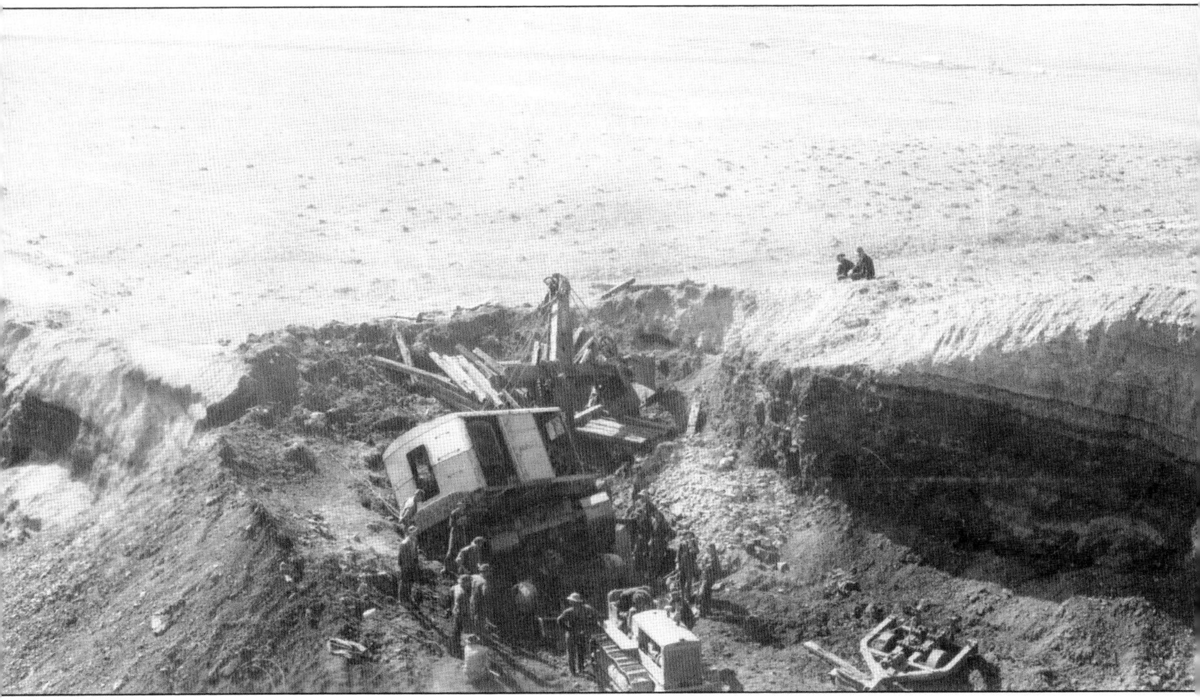

CCC workers try to upright a shovel that has tipped over while moving large timbers at a coal fire project. Though it may not have been much help, a small tractor was brought in to assist in up righting the shovel. On the Wyoming prairie in the 1930s, resources were few and far between. Sometimes the necessary equipment could be borrowed from Wyodak Mine or from a local rancher. In 1940 alone, instructors trained 42 truck drivers, 5 power shovel operators, 3 men to operate tractors, 3 road grader operators, and 190 fire fighters. Fortunately, some of the enrollees were experienced coal field workers who could teach the other men the nuances of coal mine safety. One way an underground coal fire was discovered was when the ground caved in under a rancher riding his property.

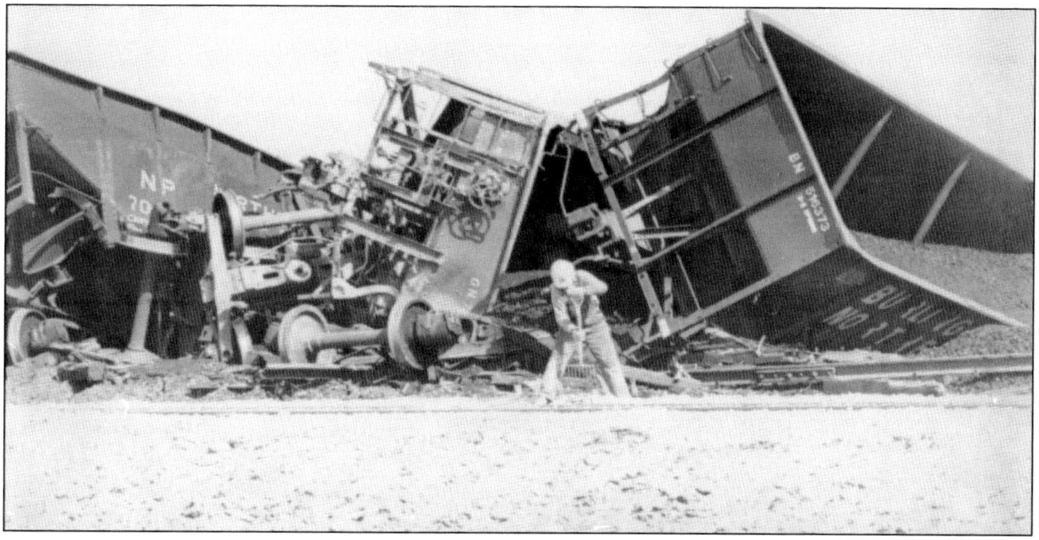

Coal cars, pulled by a Burlington Northern engine, derailed somewhere in Campbell County. Because the photograph appears to have been before the 1970s, it was most likely loaded with Wyodak coal.

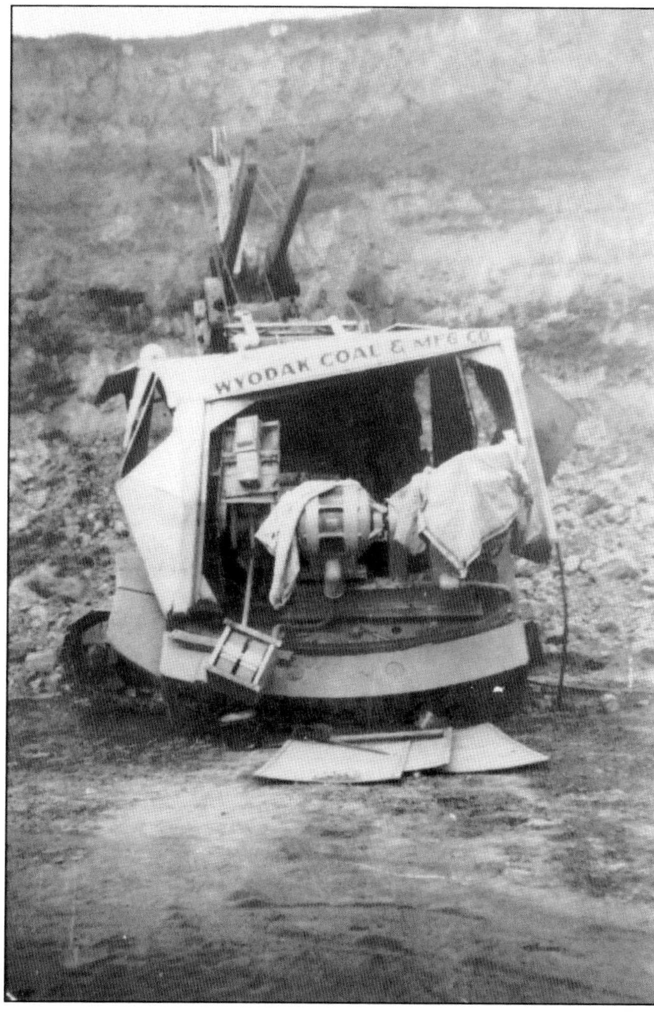

An electric shovel fell down the highwall into the pit. Although the date is uncertain, the shovel has "Wyodak Coal Company" printed over the cab.

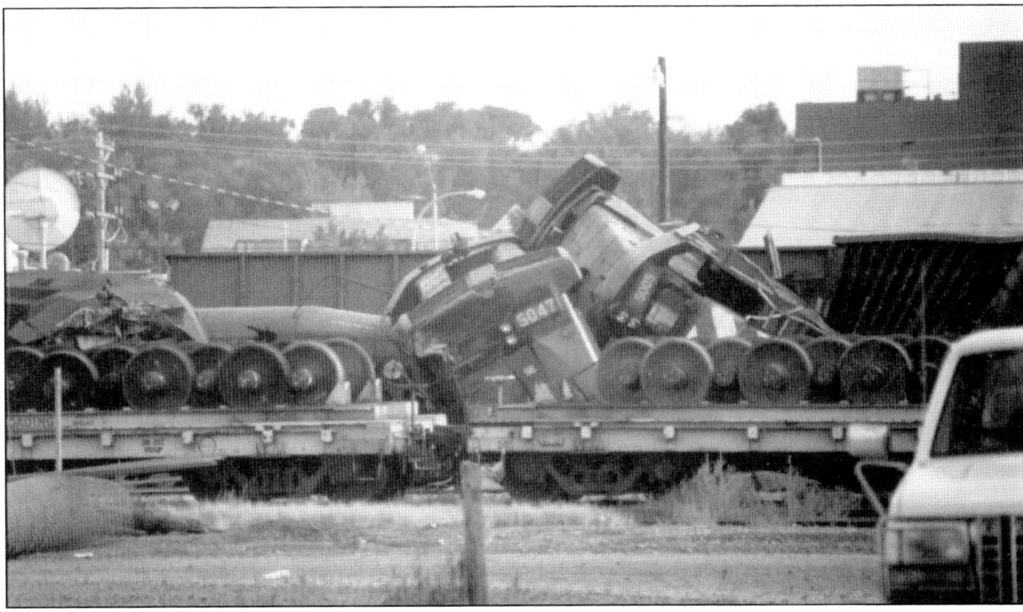

On September 7, 1994, a coal train derailed in downtown Gillette, near Lula Belle's restaurant at First Street and Gillette Avenue. Just before the collision, the conductor and engineer leaped from the train. No one was hurt, but the tracks were closed for four hours until the wreckage could be cleaned up by a derailment crew. The tall building in the background is the First National Bank.

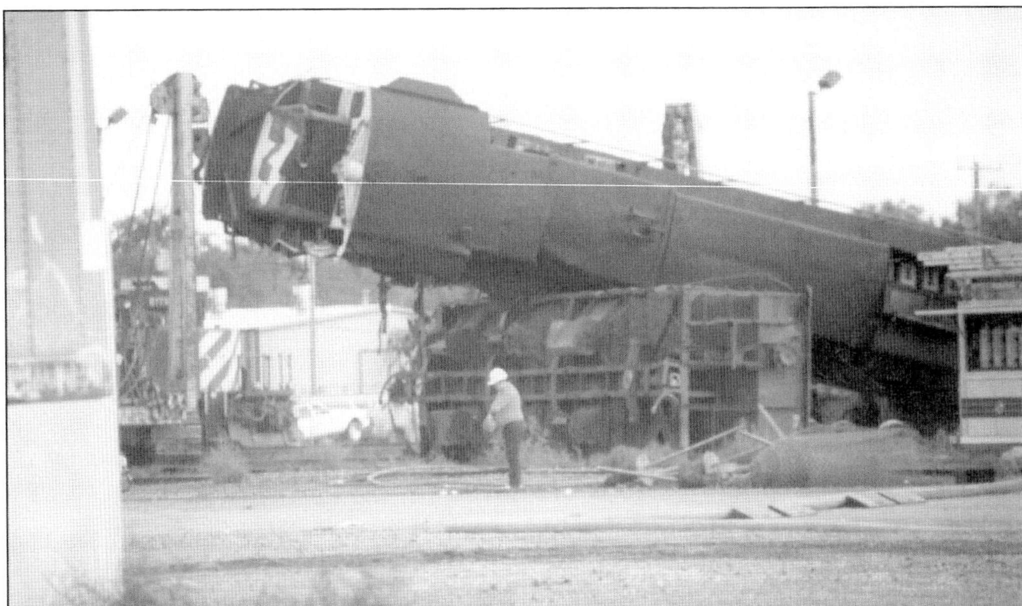

A member of the Burlington Northern derailment crew works on documenting the wreck. The locomotive rests on its side on top of a coal car. Because of the weight of a loaded coal train, two or more locomotives are often required. A second locomotive is seen here to the left of the wreckage. The fully loaded coal train crashed into a parked train below the concrete footbridge, and cars were thrown into a concrete pillar, collapsing the south end of the bridge onto the tracks.

Haul truck No. 172 overturned at Belle Ayr Mine on June 2, 1980. A large crane is pulling the truck back onto its tires. Current haul trucks with a 360-ton capacity can have a gross weight up to 1,316,000 pounds.

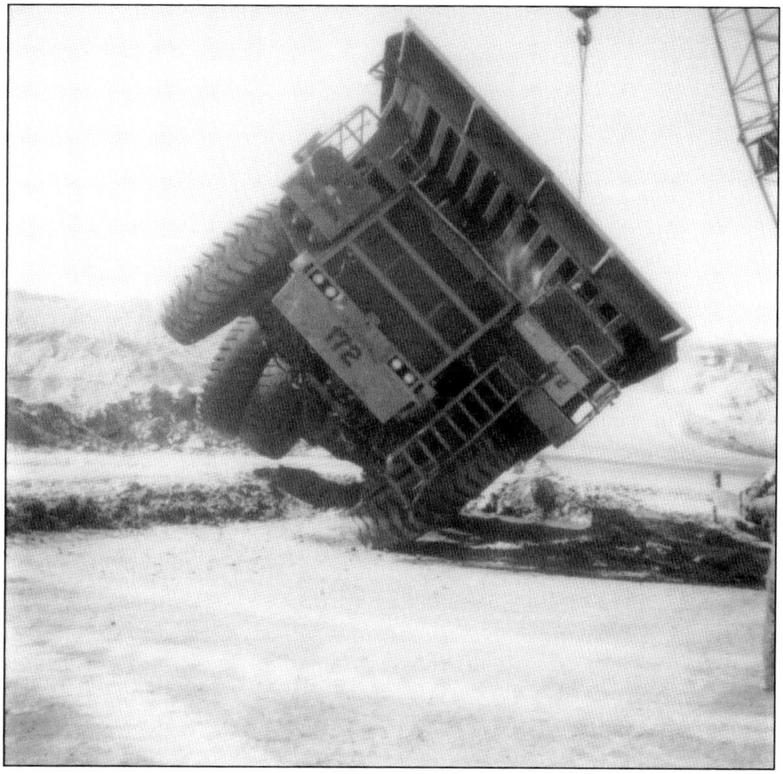

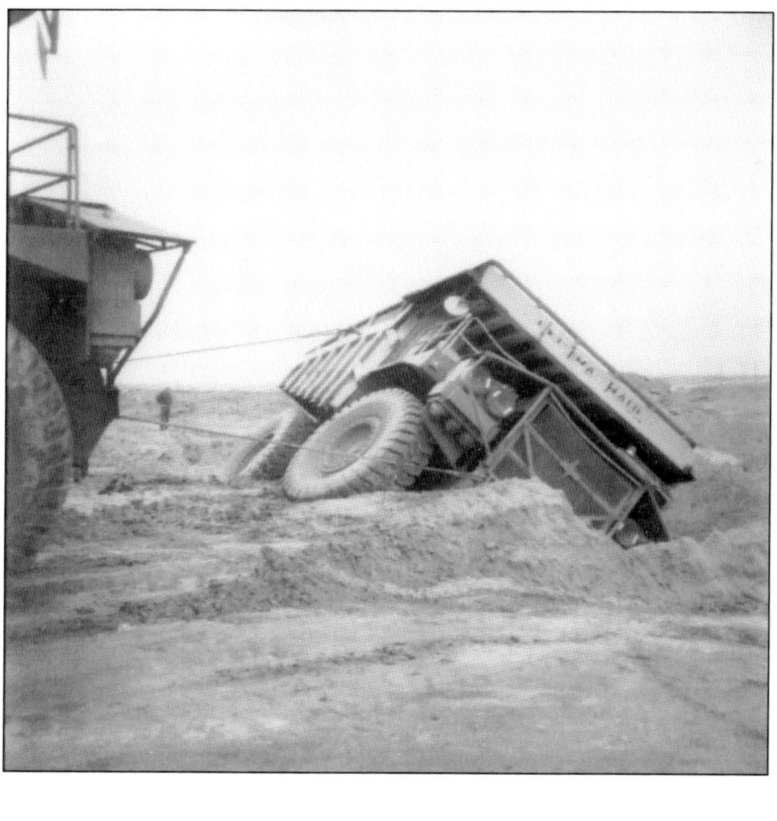

This image shows a Lectra Haul brand truck, which appears to have fallen into a sinkhole. This accident took place at Belle Ayr Mine during the 1980s. Lectra Haul is the brand name of haul trucks manufactured by Unit Rig, a company that dates back to 1935. Lectra Haul was sold to Terex, then to Bucyrus Erie, who were taken over by Caterpillar, Inc., in June 2011.

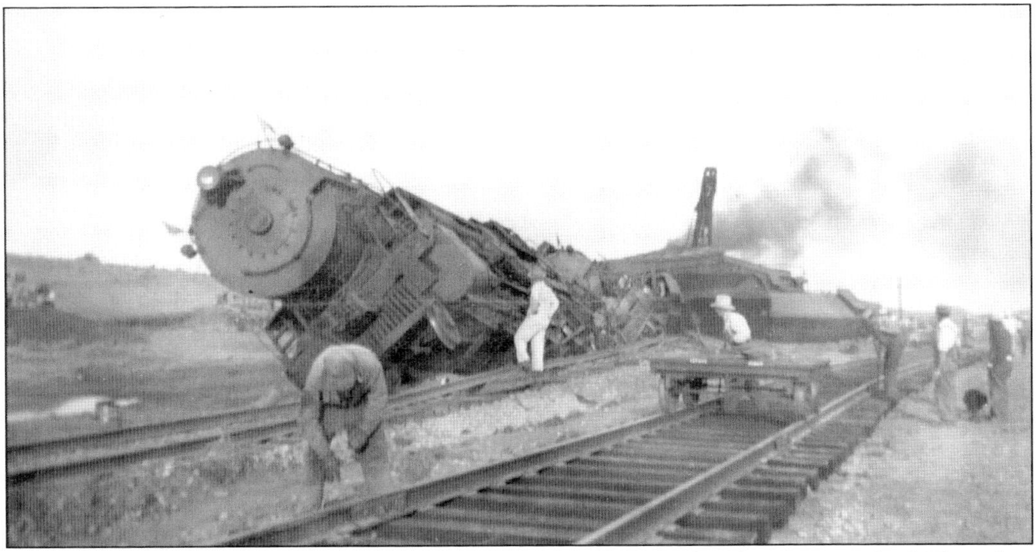

Two men were killed when a Chicago, Burlington & Quincy freight train crashed Thursday, September 11, 1952, near the Sparta siding, about six miles west of Gillette. Killed in the second engine of the double-header freight train were O.B. Burnworth, engineer, and W.J. Baas, fireman. Apparently, the head engine derailed, and the momentum of the 56-car freight piled eight cars of livestock into the rear of the second engine. (Courtesy of Fulkerson-Keck Photograph Collection.)

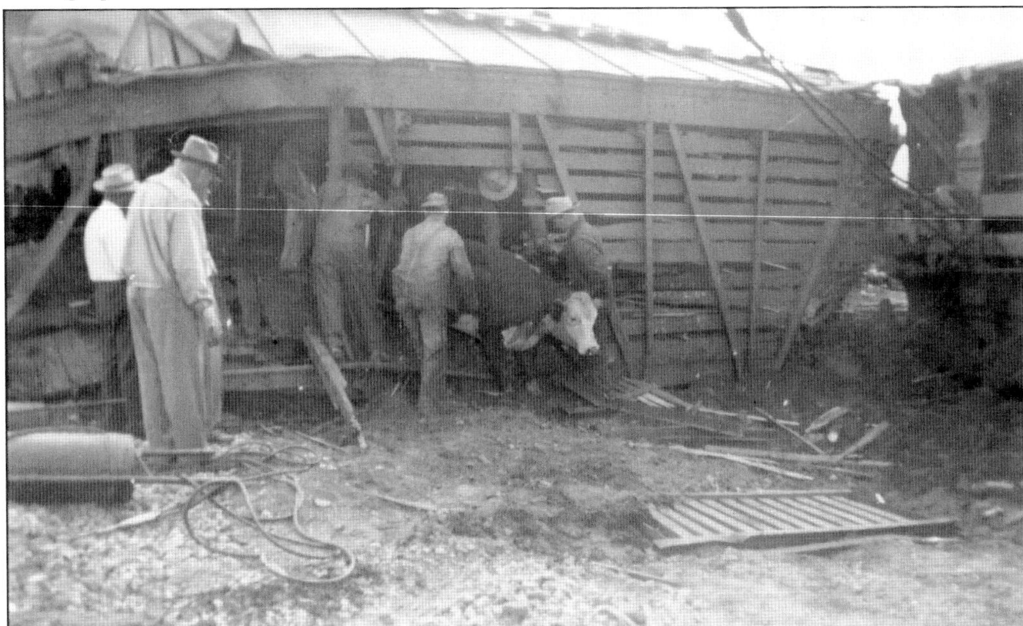

Behind the engines was the coal car to provide steam to run the train. The Peter Kiewit Sons construction company was hired to level the dirt for a new rail line so train traffic could continue, bypassing the scene of the wreck. The Peter Kiewit Sons Company, which began in 1884, would diversify over the years and become Kiewit Mining Group, the parent company of Buckskin Mining Company in Gillette, Wyoming. (Courtesy of Fulkerson-Keck Photograph Collection.)

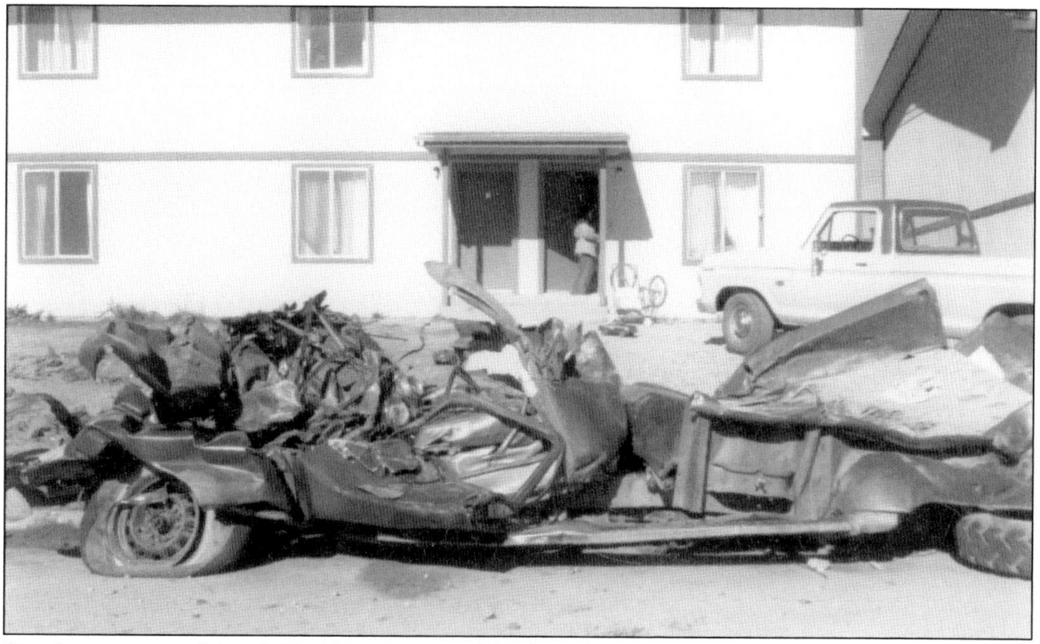

On the evening of April 30, 1981, 24-year-old John Thompson attended a going-away party for a co-worker, a foreman at Daniels Construction, where Thompson had worked as an electrician. After an evening of drinking, he "kind of blacked out" but assumed he got a ride home to Indian Hills. Around 3:00 a.m., he jumped on a D9H Caterpillar bulldozer owned by Kerr-McGee, parked near the Ramada Inn, which was at Second Street and Gurley Avenue. Kerr-McGee had loaned the 47-ton machine to the City of Gillette to help build a floodwater channel in conjunction with the Gurley Avenue railroad overpass project. (Courtesy of Lynn Stodola.)

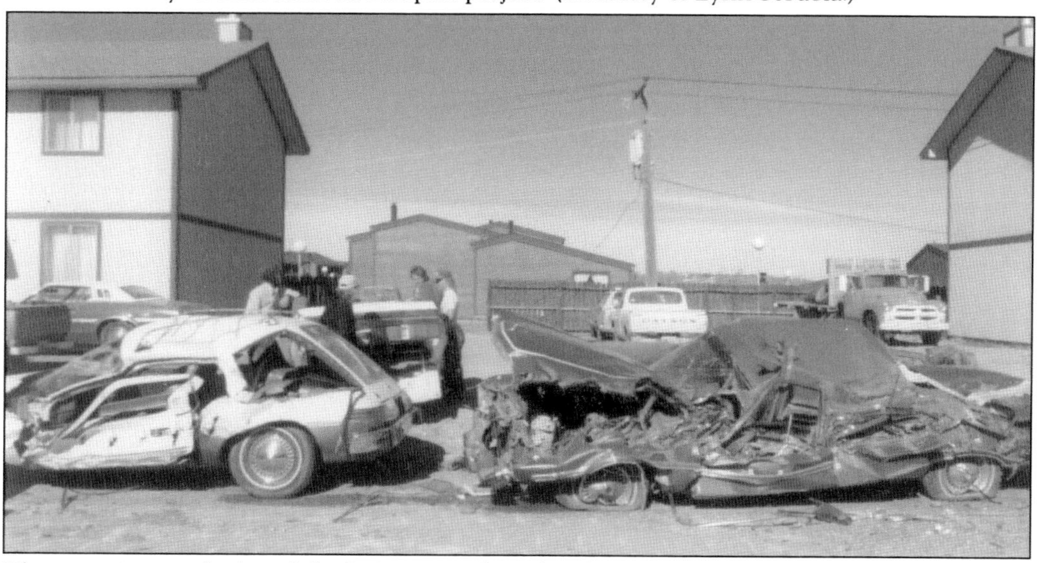

Thompson's joy ride demolished 19 cars and trucks, damaged a farm tractor and at least three multifamily dwellings, knocked down utility poles, and tore up street pavement. Initial damage estimates were between $1 million and $3 million. An actual total was considered "impossible to pinpoint." During the melee, all 29 Gillette police officers, including Chief Don Schneider, were called out to the scene. (Courtesy of Lynn Stodola.)

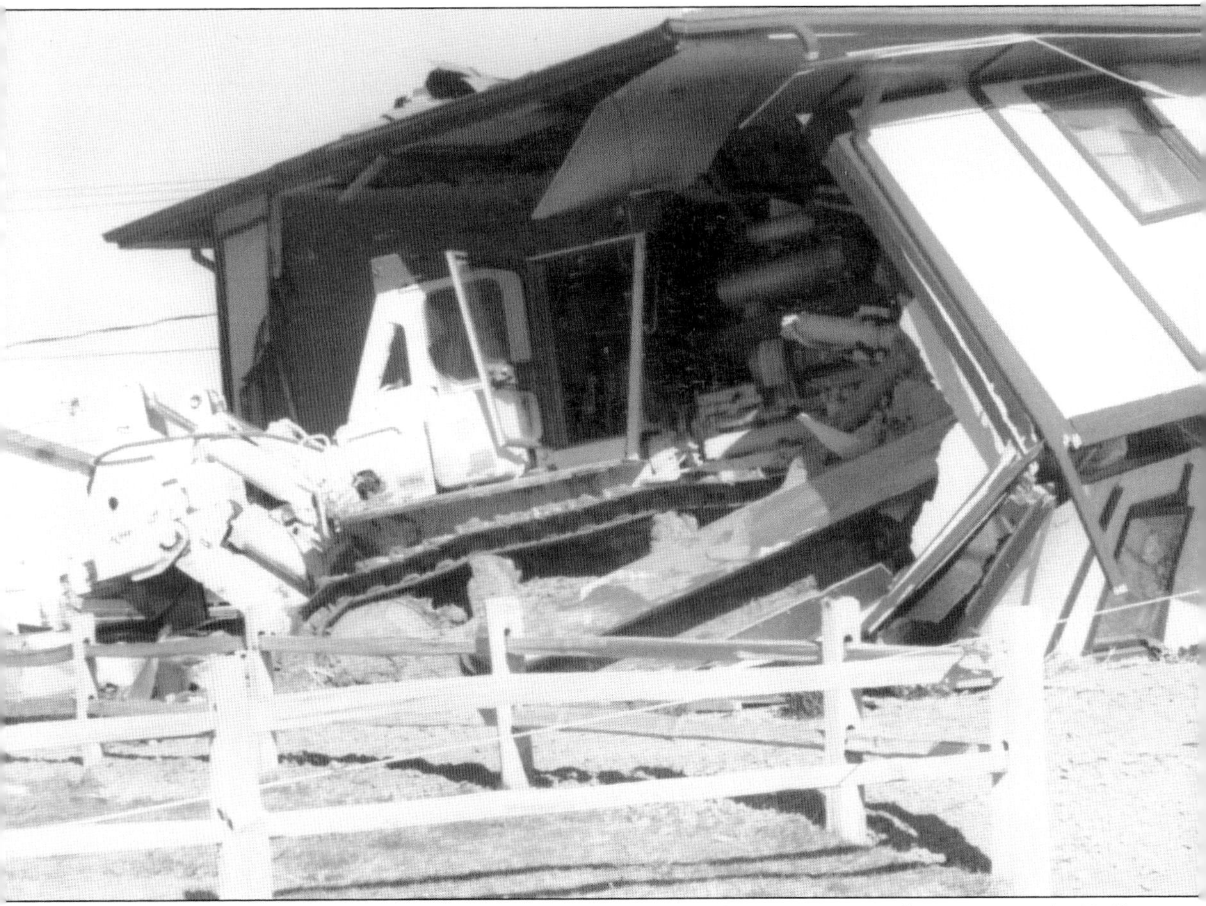

Around 3:45 a.m., Patrolman David Adsit shot four slugs from a 12-gauge shotgun into the D-9's engine, causing it to smoke. Whenever policemen came within a quarter of a block of the bulldozer, it would spin around and come after them. The bulldozer proceeded to break a gas line, knock down more power lines, and headed toward some apartment buildings. Police thought they might have to kill the driver before he killed someone else. It was about this time that Thompson jumped from the bulldozer, which then ripped through the side of an apartment building at 1203 East Tenth Street and entered the side of another building at 904 Church Street. Thompson fell asleep in a nearby apartment, and when he woke up he went outside and saw the damage. He thought a tornado had come through until he saw the tracks. (Courtesy of Lynn Stodola.)

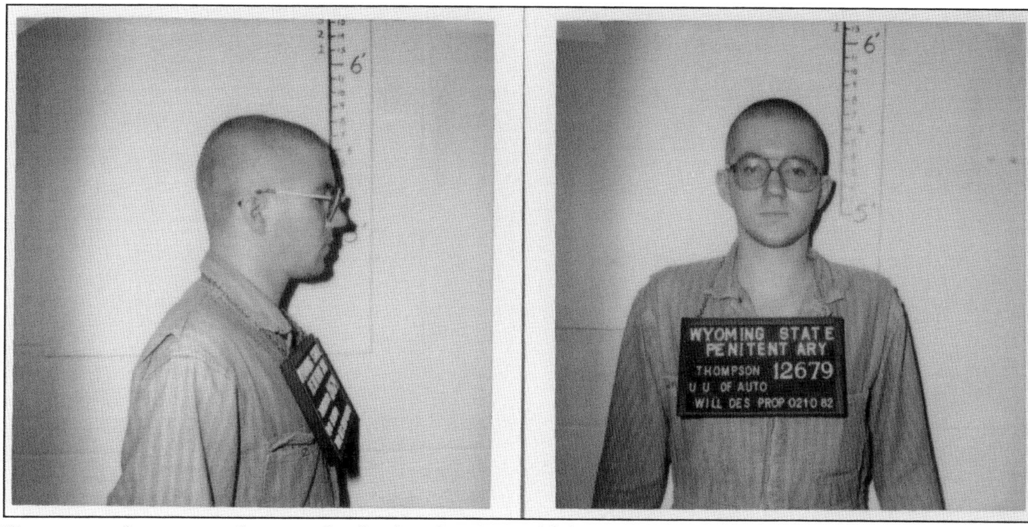

"It surprised me more than anybody else that I would do something like that" said John Thompson in February 1982, just minutes after he was sentenced to up to 10 years in prison on charges of felony destruction of property and vehicle theft. Thompson attributed the rampage to drinking alcoholic beverages, which he had been told causes some people to lose nearly all self-control.

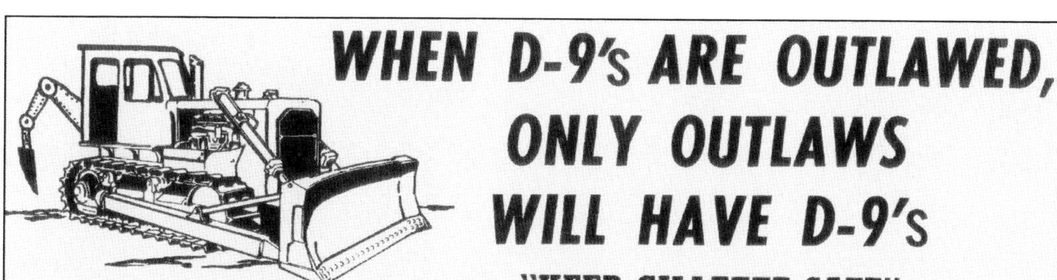

Newspapers around the world poked fun at Gillette after the incident. The *Daily Mail* in London wrote, "Unfortunately for Thompson, one of the few buildings left standing was the town jail." Gillette mayor Mike Enzi received "The Wyoming Association of Municipalities First Annual Urban Renewal Award." T-shirts and bumper stickers appeared for sale in Gillette stores. Kerr-McGee Coal Company denied any responsibility for the event or the damage.

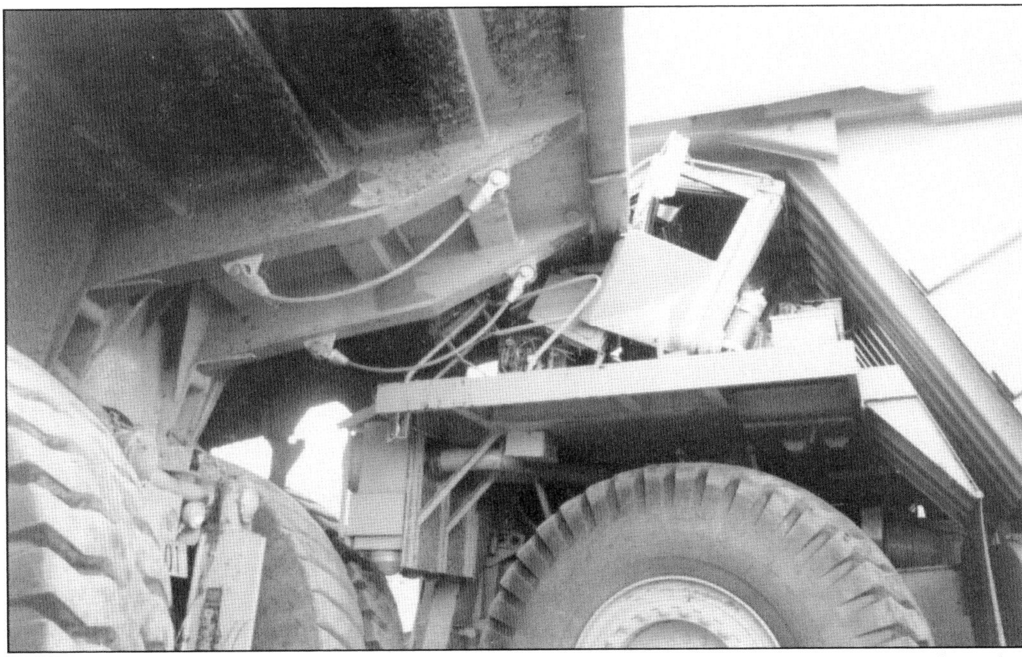

Rawhide Mine was the location of a wreck between two 170-ton haul trucks. One truck made a quick U-turn in front of another loaded haul truck. The radio microphone in the first truck fell from its spot on the ceiling and hit the parking brake in the cab, stopping the truck immediately. The loaded haul truck could not stop in time and ran into the first truck. (Courtesy of Keith Fare.)

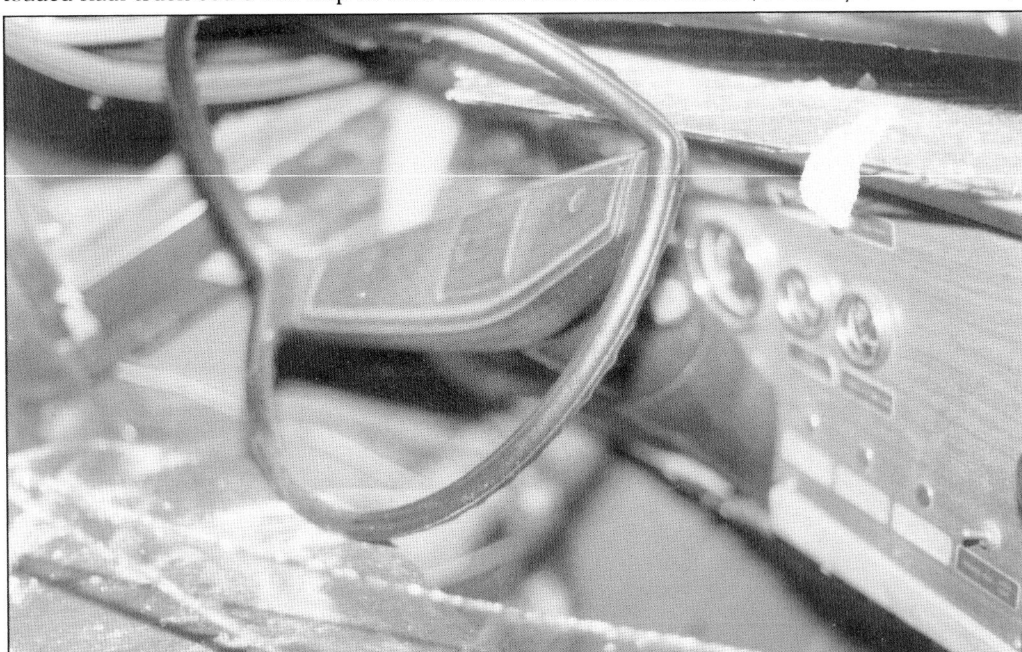

If the loaded truck had swerved to the left, it would have hit a Suburban. The driver of the loaded truck escaped injury by diving under the steering wheel. As a result of that accident, radio microphones were moved down to a lower spot in the cab the next day. (Courtesy of Keith Fare.)

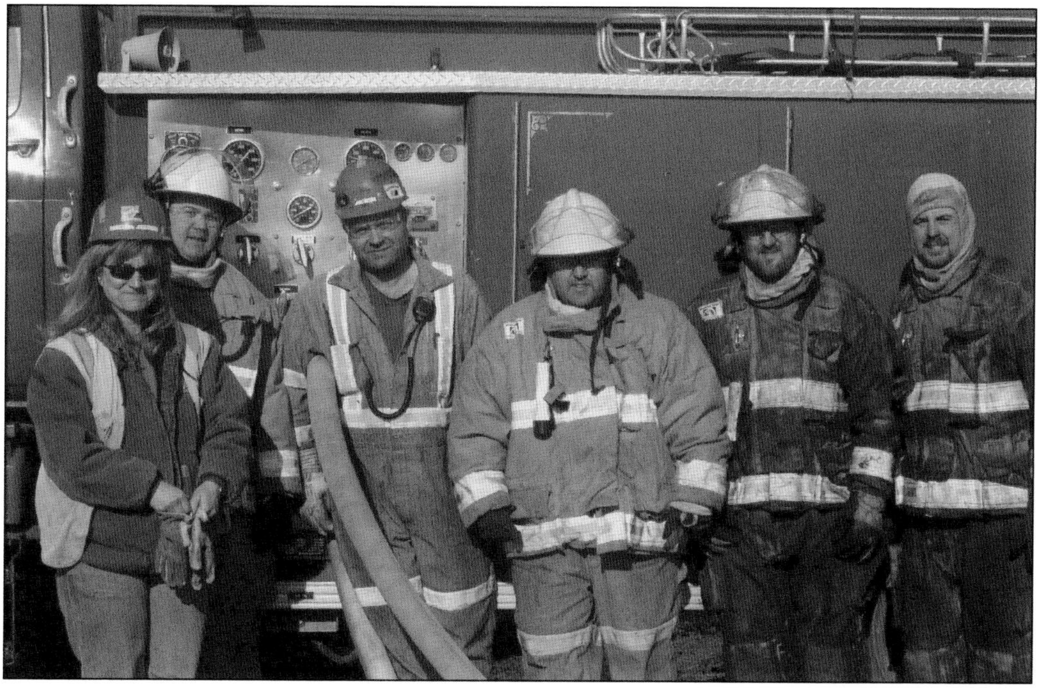

Rawhide Mine Rescue personnel pause for a group photograph after extinguishing a transfer fire at Rawhide Mine. Shown from left to right are Mary DeRudder, Andy Pond, Brandon Covell, Tom Gorton, Rick Stainbrook, and Tim Colombatto. (Courtesy of Campbell County Fire Department.)

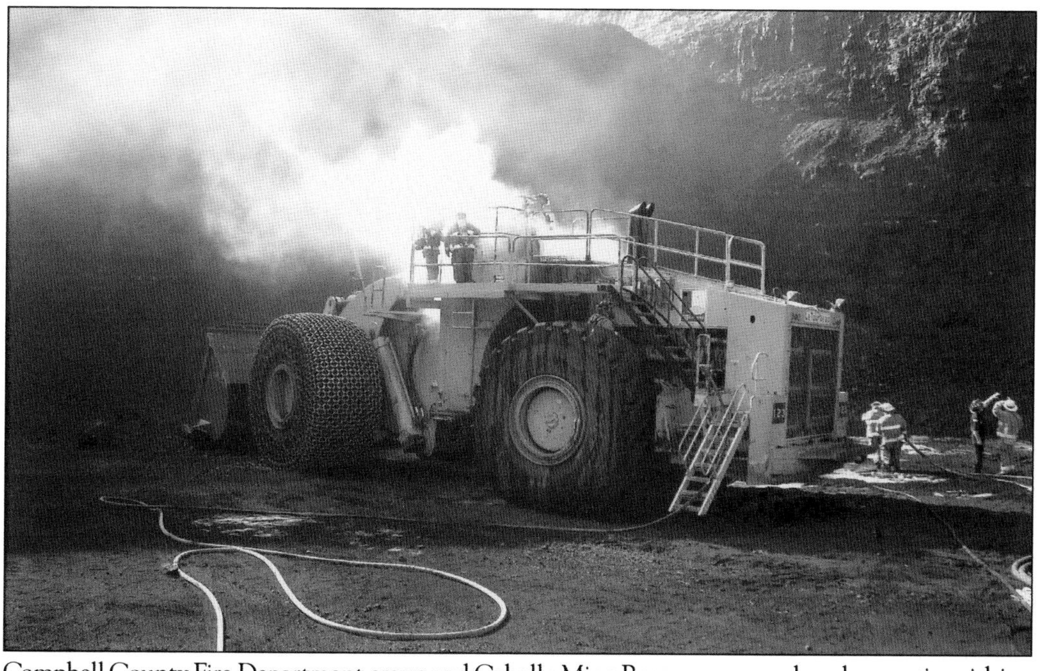

Campbell County Fire Department crews and Caballo Mine Rescue personnel work on extinguishing a loader fire at Caballo Mine in March 2010. Campbell County Fire Department has been partly paid staff and partly volunteer for many years. (Courtesy of Campbell County Fire Department.)

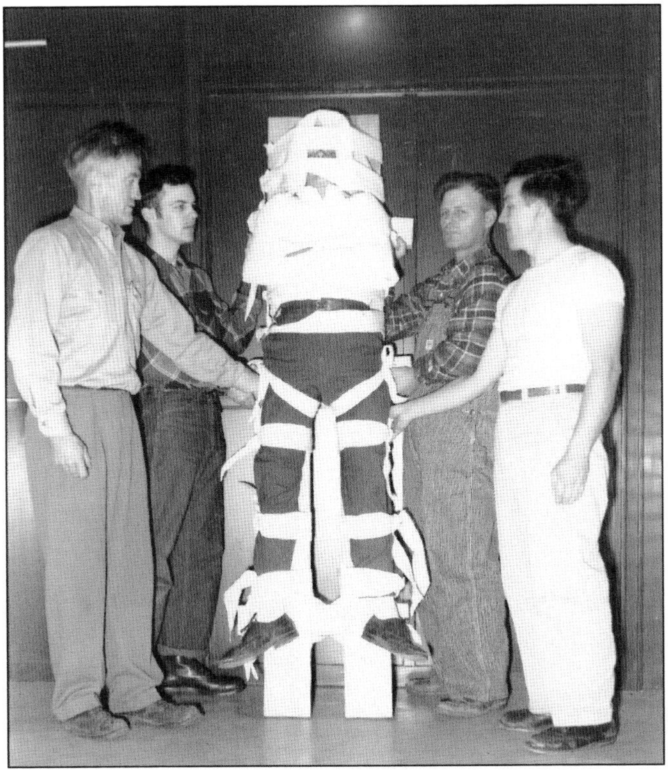

Mine rescue training at Homestake required a commitment from the "victim." Everyone was expected to know basic first-aid techniques, since emergency personnel were not as readily available as they are today. Mine rescue training for fires, explosions, and cave-ins dates back to the beginning of the 20th century, before the United States Bureau of Mines was created in 1910. (Courtesy of Homestake Adams Research and Cultural Center.)

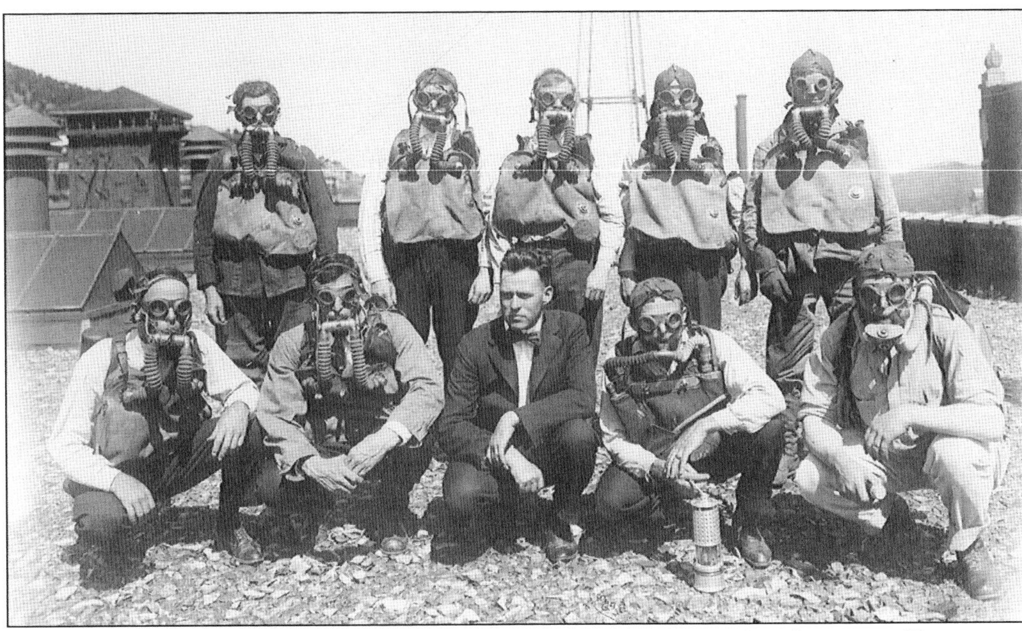

Homestake rescue personnel pose with oxygen breathing apparatus. A closed-circuit, self-contained breathing apparatus was a device used during underground mine rescues where narrow spaces were often too tight a squeeze for an apparatus with oxygen tanks. Most mine rescue teams are composed of miners who know the particular mine, the type of equipment they may encounter, and the geological conditions. (Courtesy of Homestake Adams Research and Cultural Center.)

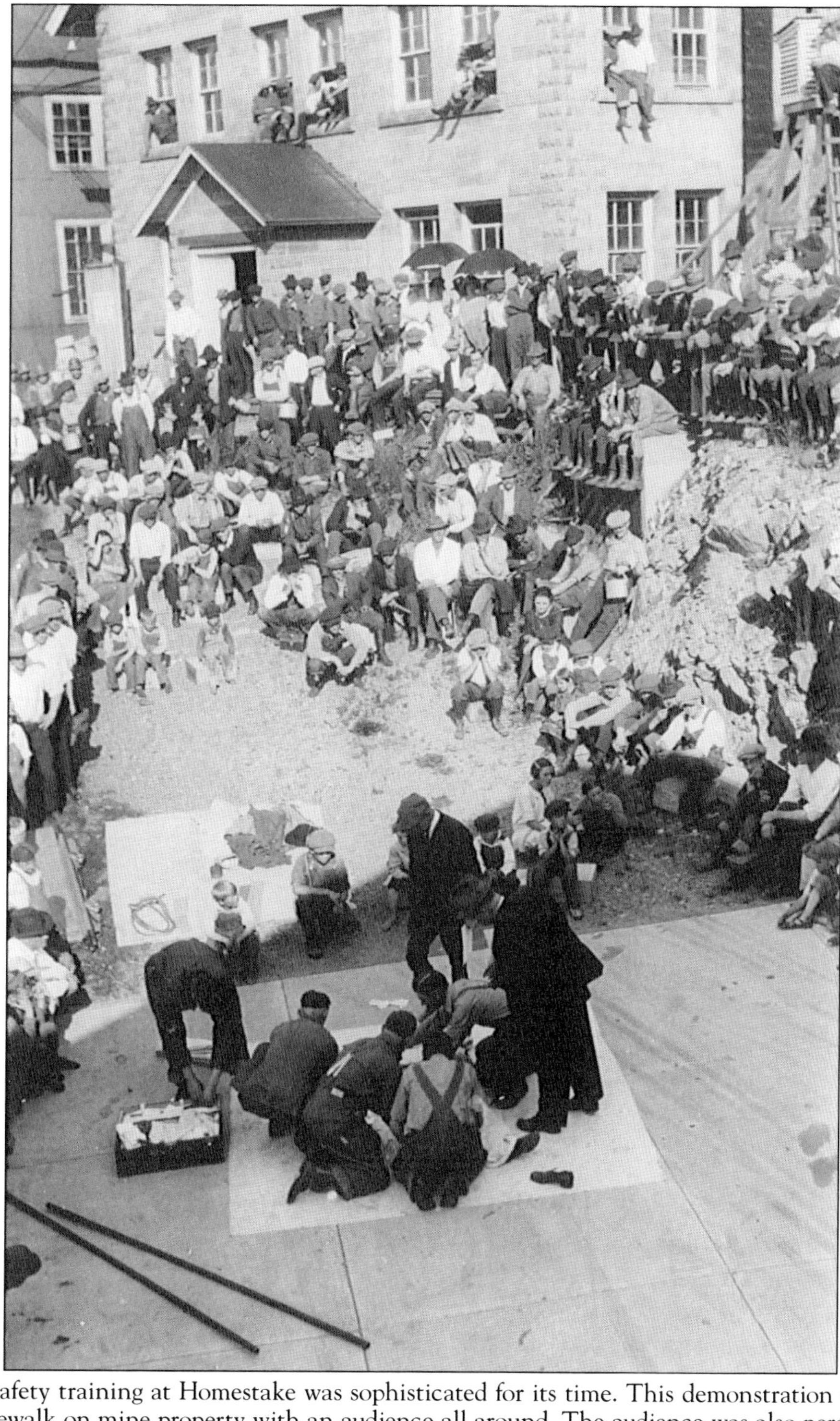

Mine safety training at Homestake was sophisticated for its time. This demonstration was on the sidewalk on mine property with an audience all around. The audience was also perched in open windows on the second floor, not exactly safety-conscious. (Courtesy of Homestake Adams Research and Cultural Center.)

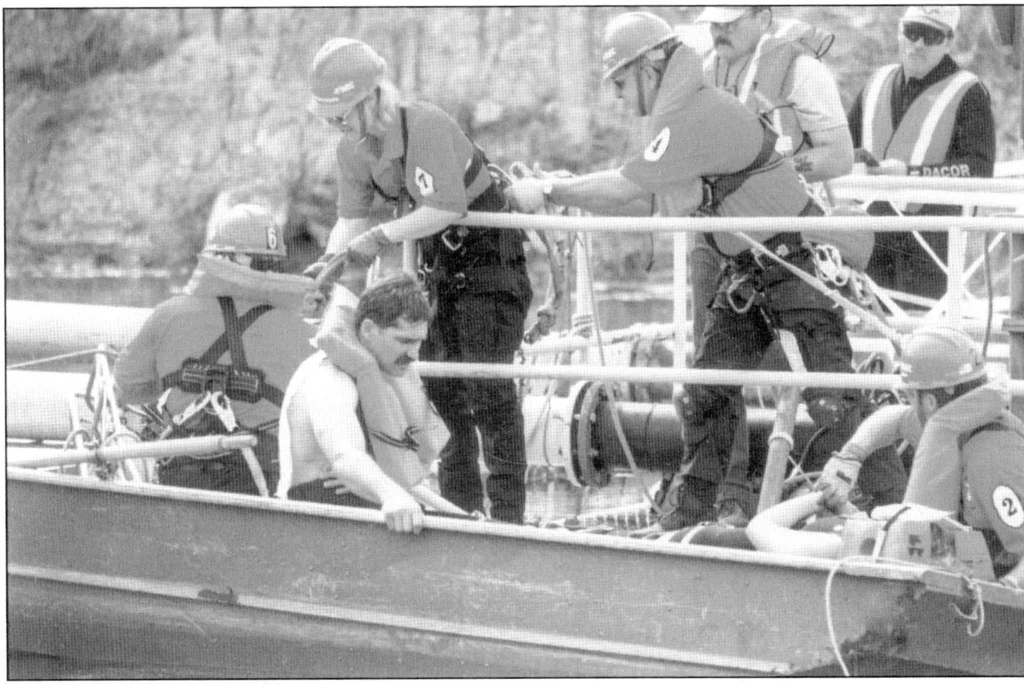

Mine rescue teams in the Powder River Basin started training in 1979. A committee visited Canada to learn from their established mine rescue competition. In 1980, Rawhide and Caballo Mines held their own competition, and, by 1981, the competition grew larger. In this scenario, the valve was knocked off an oxygen/acetylene cylinder. The falling cylinder knocked Lyn Busskohl (with his back to camera) into the water. (Courtesy of Sandie Jones.)

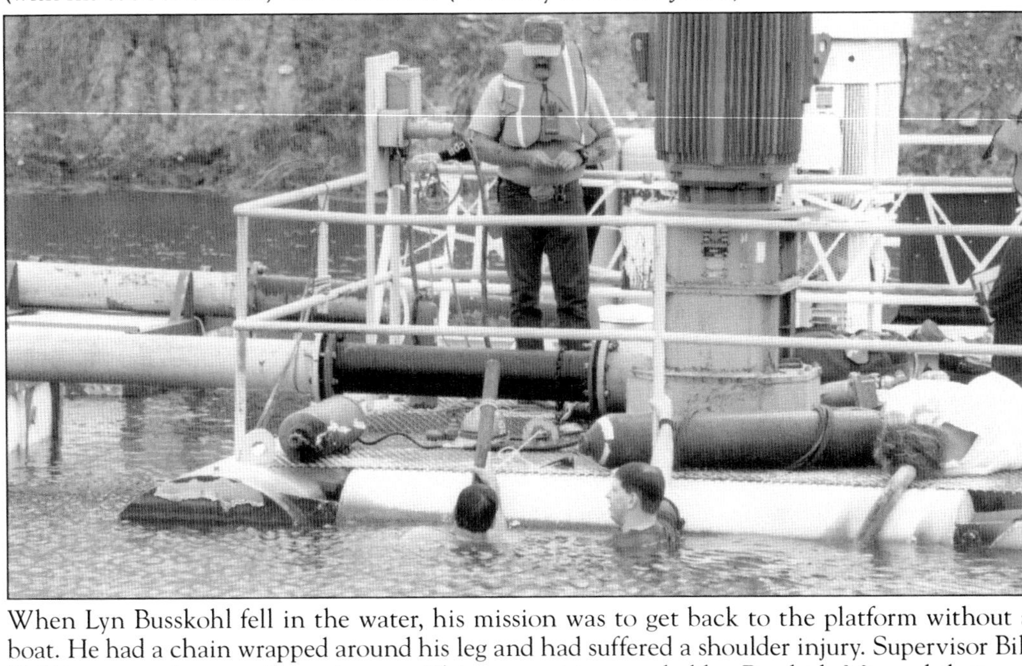

When Lyn Busskohl fell in the water, his mission was to get back to the platform without a boat. He had a chain wrapped around his leg and had suffered a shoulder injury. Supervisor Bill Hoblit swam to Lyn to offer assistance. This competition was held at Rawhide Mine while it was temporarily idle. (Courtesy of Sandie Jones.)

Mine rescue competition belt buckles depict a scene from the previous year's competition. Belt buckles were given as prizes starting in 1985. Prior to that, prizes were plaques. Competition is limited to 12 teams. Our Powder River Basin competition includes teams from Mexico, Australia, Indonesia, and Canada, in addition to the United States.

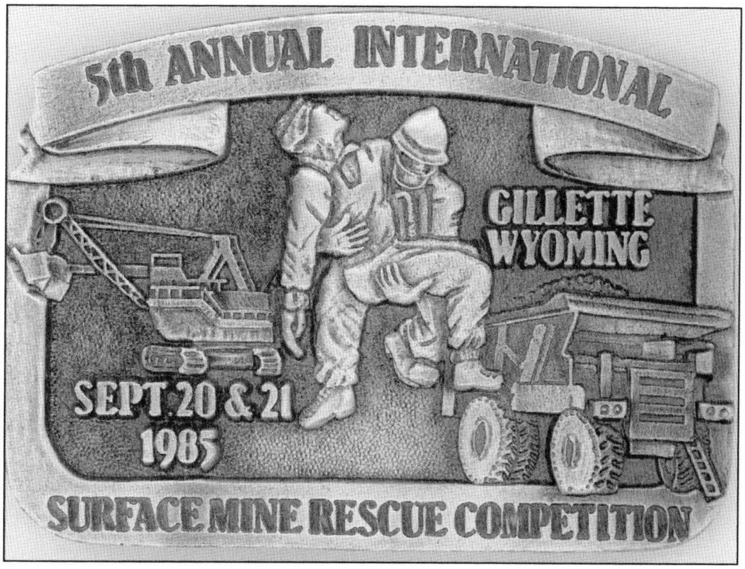

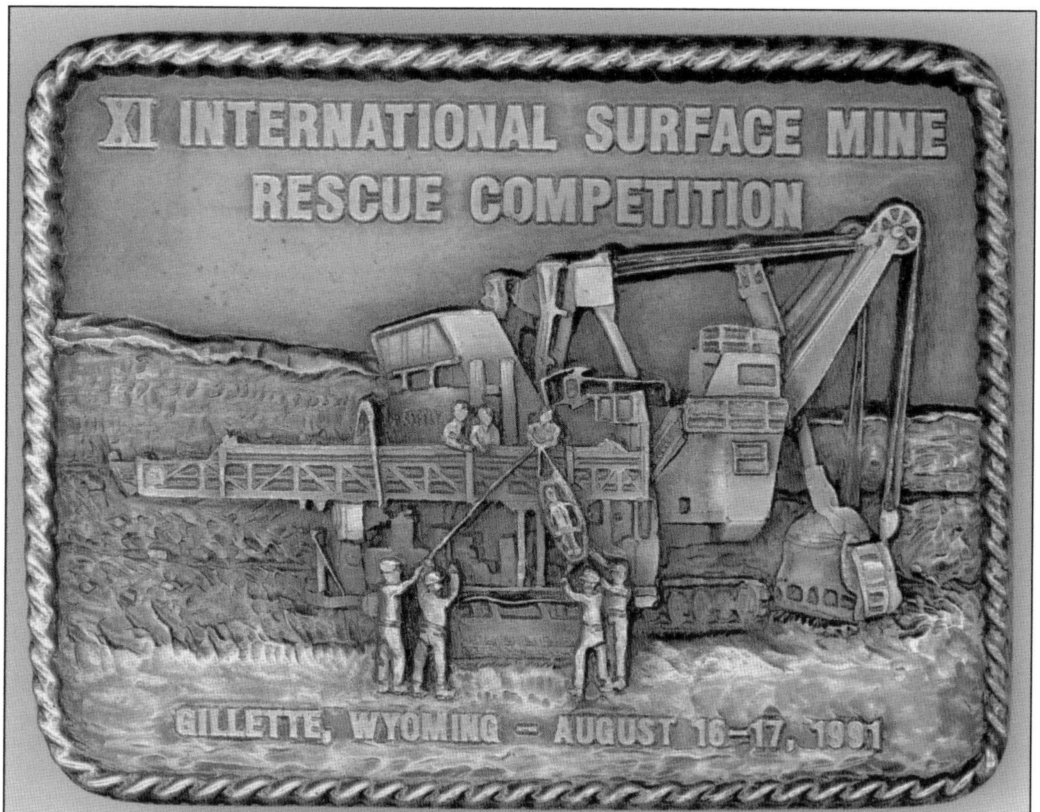

The host team is responsible for creating the rescue scenario, which often comes from a real event. The early competition was only one day; now it is a two- to three-day competition that includes one full day of training. The host could be a coal mine or a business that provides products and/or services to the mining industry. The 32nd-annual International Surface Mine Rescue Competition in 2012 was hosted by Joy Global.

BIBLIOGRAPHY

Black Hills Pioneer. *Homestake 1876–2001*. December 2001.

Buchsbaum, Lee. "Cloud Peak Energy Prepares for the Long Run." *Coal Age*. June 2011.

Civilian Conservation Corps. Consolidated Report for 1933–1938.

Campbell County Historical Society. *The Treasured Years*. Walsworth Publishing Company, 1991.

Considine, Timothy J. "Powder River Basin Coal: Powering America, Final Report to the Wyoming Mining Association." Wyoming Mining Association. December 21, 2009.

Gallagher, Megan M. *Peabody Energy–Celebrating 125 Years*. Peabody Energy: 2008.

Garrett, Lewis R., Black Thunder report. April 1989.

Homestake Mining Company. *Sharp Bits*. October 1955.

Lynde, Marilyn J., *The Life and Sacred Times of Timber Jack Joe*. Casper, WY: Mountain States Lithographing, October 2000.

Wyoming Mining Association. *The 2011 Concise Guide to Wyoming Coal*. Cheyenne, WY: August 2011.

INDEX